Calculus&Mathematica

APPROXIMATIONS:
Measuring Nearness

Bill Davis
Ohio State University

Horacio Porta
University of Illinois, Urbana-Champaign

Jerry Uhl
University of Illinois, Urbana-Champaign

Addison-Wesley Publishing Company

Reading, Massachusetts • Menlo Park, California • New York
Don Mills, Ontario • Wokingham, England • Amsterdam • Bonn
Sydney • Singapore • Tokyo • Madrid • San Juan • Milan • Paris

None of Addison-Wesley, Wolfram Research, Inc., or the authors of Calculus&*Mathematica* makes any warranty or representation, either express or implied, with respect to the Calculus&*Mathematica* package, or its performance, merchantability or fitness for particular purposes.

None of Addison-Wesley, Wolfram Research, Inc., or the authors of Calculus&*Mathematica* shall be liable for any direct, indirect, special, incidental, or consequential damages resulting from any defect in the Calculus&*Mathematica* package, even if they have been advised of the possibility of such damages. Some states do not allow the exclusion or limitation of implied warranties or consequential damages, so the above exclusion may not apply to you.

Mathematica is a registered trademark, and *MathLink* and *MathReader* are trademarks of Wolfram Research, Inc.

Mathematica is not associated with Mathematica, Inc., Mathematica Policy Research, Inc., or MathTech, Inc.

Many of the designations used by manufacturers and sellers to distinguish their products are claimed as trademarks. Where those designations appear in this book, and Addison-Wesley was aware of a trademark claim, the designations have been printed in initial cap or all caps.

Copyright © 1994 by Addison-Wesley Publishing Company, Inc.

All rights reserved. No part of this publication may be reproduced, stored in a retrieval system, or transmitted, in any form or by any means, electronic, mechanical, photocopying, recording or otherwise, without the prior written permission of the publisher. Printed in the United States of America.

ISBN 0-201-58468-9

1 2 3 4 5 6 7 8 9-CRS-98 97 96 95 94 93

Preface

Did you ever wonder how to calculate truly accurate values of sin[x] or cos[x]? One way to get a rough estimate of the values of sin[x] and cos[x] is to draw the appropriate right triangle and to make the measurements of opposite/hypotenuse and adjacent/hypotenuse, but you cannot expect to get truly accurate estimates this way.

Did you ever wonder how to calculate accurate values of e^x? When you think about it, you'll probably agree that you can estimate $e^{1/2}$ by approximating e by, say, 2.718281828 and then taking the square root of 2.718281828. This will give you a reasonably good estimate of $e^{1/2}$ and if you need a better approximation of $e^{1/2}$, you can use more accurate decimals of e. When you try to calculate at e^π, things get squirrelly because there is no physical process (like taking square roots) that corresponds to the calculation of e^π.

Did you ever wonder how to calculate accurate values of log[x]? This is a real problem because e is the base of log[x] and you'll never write down a completely accurate decimal number that equals e. What do you? Just give up? Heck no!

Instead of trying to calculate sin[x], cos[x], e^x, and log[x] directly, you'll learn how to calculate them by approximating them. For instance, you'll see why it's natural that when you want to calculate e^x for a reasonably small number x, you can get a pretty good estimate by calculating

$$1 + x + \frac{x^2}{2!},$$

and you can improve your estimate by calculating

$$1 + x + \frac{x^2}{2!} + \frac{x^3}{3!}.$$

Or, if you want even more accuracy, you'll see why it's natural to go with
$$1 + x + \frac{x^2}{2!} + \frac{x^3}{3!} + \frac{x^4}{4!} + \frac{x^5}{5!} + \frac{x^6}{6!},$$
or even
$$1 + x + \frac{x^2}{2!} + \frac{x^3}{3!} + \frac{x^4}{4!} + \frac{x^5}{5!} + \frac{x^6}{6!} + \frac{x^7}{7!} + \frac{x^8}{8!} + \frac{x^9}{9!}.$$

The approximating schemes you'll learn are the approximation schemes on which calculators and *Mathematica* base their calculations of $\sin[x]$, $\cos[x]$, e^x, and $\log[x]$, as well as many other functions.

Along the way you'll see how these approximations are based on the same idea that highway engineers use to merge two roads smoothly. And you'll see how to use these approximations for automatic landing of an airplane, how to slam out accurate decimals of π, how to calculate some limits, how to use Issac Newton's root-finding method, how to prescribe dosing of drugs, and how to resolve the controversy between parabolic and spherical reflectors.

If you're going on in math, physics, or engineering, you'll want to know about these approximations because you'll see them again and again.

How to Use This Book

In Calculus&*Mathematica*, great care has been taken to put you in a position to learn visually. Instead of forcing you to attempt to learn by memorizing impenetrable jargon, you will be put in the position in which you will experience mathematics by seeing it happen and by making it happen. And you'll often be asked to describe what you see. When you do this, you'll be engaging in active mathematics as opposed to the passive mathematics you were probably asked to do in most other math courses. In order to take full advantage of this crucial aspect of Calculus&*Mathematica*, your first exposure to a new idea should be on the live computer screen where you can interact with the electronic text to your own satisfaction. This means that you should avoid "introductory lectures" and you should avoid reading this book at first. After you have some familiarity with new ideas as found on the computer screen, you should seek out others for discussion and you can refer to this book to brush up on a point or two after you leave the computer. In the final analysis, this book is nothing more than a partial record of what happens on the screen.

Once you have participated in the mathematics and science of each lesson, you can sharpen your hand skills and check up on your calculus literacy by trying the questions in the Literacy Sheet associated with each lesson. The Literacy Sheets appear at the end of each book.

Significant Changes from the Traditional Course

Student-produced splines are used for motivation.

Series of numbers are deemphasized in favor of using series of functions for approximation.

For the first time, the complex singularity criterion for convergence of expansions is revealed to calculus students.

Taylor's formula comes near the end, where it can be appreciated instead of at the beginning.

Students find out what expansions are and what they are good for by visualization instead of hearing about them in a lecture.

Convergence of expansions at the "endpoints" is not treated.

The traditional collection of convergence tests is gone.

Limits of quotients at a point are calculated by dividing the leading term of the expansion of the denominator into the leading term of the expansion of the numerator—just the way most scientists and mathematicians do this.

L'Hospital's rule is put in its proper calculus context as an easy consequence of Taylor's formula.

Treatment of functions defined by a power series via a differential equation makes it possible to eliminate this topic from follow-up courses in differential equations.

Contents

APPROXIMATIONS: Measuring Nearness

In normal use, the student engages in all the mathematics and the student engages in selected experiences in math and science as assigned by the individual instructor.

4.01 Splines   1

Mathematics Remarkable plots explained by order of contact. Splining for smoothness at the knots.

Science and math experience Experiments geared at discovering that the smoother the transition from one curve to another at a knot, the better both curves approximate each other near the knot. Splining functions and polynomials. Splines in road design. Landing an airplane. The natural cubic spline. Order of contact for derivatives and integrals.

4.02 Expansions in Powers of x 29

Mathematics The expansion of a function $f[x]$ in powers of x as a file of polynomials with higher and higher orders of contact with $f[x]$ at $x = 0$. The expansions every literate calculus person knows ($1/(1-x)$, e^x, $\sin[x]$, and $\cos[x]$). Expansions for approximations.

Science and math experience Experiments geared toward discovering that using more and more of the expansion results in better and better approximation. Halley's way of calculating accurate decimals of π. Expansions by substitution. Expansions by differentiation. Expansions by integration. Recognition of expansions. Expansions that satisfy a priori error bounds.

4.03 Using Expansions   57

Mathematics The expansion of a function $f[x]$ in powers of $(x - b)$ as a file of polynomials with higher and higher orders of contact with $f[x]$ at $x = b$. Netwon's method. Multiplying and dividing expansions. Using expansions to help calculate limits at a point. Expansions and the complex exponential function. Using expansions to help to get precise estimates of some integrals.

Science and math experience Centering expansions for good approximation. Newton's method for root finding. Successes and failures of Newton's method. Using the complex exponential to generate trigonometric identities. Comparing reflecting properties of spherical mirrors and the reflecting properties of parabolic mirrors. Using expansions to see why spherical mirrors do have limited ability to concentrate light rays. Behavior of expansions very close to 0. Behavior of expansions far away from 0.

4.04 Taylor's Formula   101

Mathematics Taylor's formula for expansions in powers of $(x - b)$.

Science and math experience Euler, midpoint, and Runge approximations of $f[x]$ given $f'[x]$. Experiments comparing the quality of midpoint, and Runge approximations. Adaption of Euler, midpoint, and Runge approximations to approximating the plots of the differential equation $y'[x] = f[x, y[x]]$ with $y[a] = b$. Taylor's formula in reverse. L'Hospital's rule by dividing the leading term of the expansion of the denominator into the leading term of the expansion of the numerator. Centering the expansion for best approximation. Experiments comparing the derivative of the expansion and the expansion of the derivative.

4.05 Barriers to Convergence   141

Mathematics Barriers and complex singularities. The convergence interval of an expansion as the interval between the barriers. Why some functions like $1/(1+x^2)$ have barriers and others like e^x and $\sin[x]$ do not. Why functions like $x^{5/2}$ do not have expansions in powers of x but do have expansions in powers of $(x - b)$ for $b > 0$. Why the barriers for $f[x]$, $f'[x]$, and $\int_a^x f[t]\,dt$ are the same.

Science and math experience Shortcuts based on the expansion of $1/(1-x)$ in powers of x. Using the expansion of $1/(1-x)$ in powers of x for drug dosing. Infinite sums of numbers resulting from expansions. Barriers resulting from splines. Infinite sums and decimals. Experiments relating expansions in powers of x to interpolating polynomials. Runge's disaster.

4.06 Power Series 183

Mathematics Functions defined by a power series. Functions defined by power series via differential equations. The power series convergence principle, which says that if for some positive number r the infinite list

$$\{a_0, a_1 r, a_2 r^2, a_3 r^3, \ldots, a_k r^k, \ldots\}$$

is bounded, then the power series

$$a_0 + a_1 x + a_2 x^2 + a_3 x^3 + \cdots + a_k x^k + \cdots$$

converges for $-r < x < r$.

Science and math experience Experiments in trying to plot functions defined by power series. Experiments in plotting a function defined by a power series via a differential equation versus plotting the same function directly through *Mathematica*'s numerical differential equation solver. The ratio test for power series as a consequence of the power series convergence principle. The functions e^x, sin[x], and cos[x] from the viewpoint of power series. Experiments in truncation of power series. The Airy function as a function defined by a power series.

LESSON 4.01

Splines

Basics

■ B.1) Some remarkable plots explained by order of contact

Sometimes functions whose formulas are strikingly different have plots that are strikingly similar. Here are a few for your plotting pleasure.

B.1.a.i) Here are plots of $f[x] = \cos[2x]$, and $g[x] = \sqrt{1 - 4x^2}$ on the same axes for $-0.5 \le x \le 0.5$.

```
In[1]:=
  Clear[x,f,g]
  f[x_] = Cos[2 x];
  g[x_] = Sqrt[1 - 4 x^2];
  Plot[{f[x],g[x]},{x,-0.5,0.5},
  AxesLabel->{"x",""},PlotStyle->
  {{GrayLevel[0.4],Thickness[0.01]},
  {Red,Thickness[0.01]}}];
```

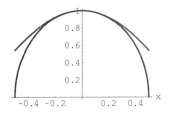

Describe what you see.

Answer: Very similar plots near $x = 0$. As you advance x from the left of 0 to the right of 0, you can make the transition from one curve to the other with ease.

B.1.a.ii) Here are plots of $f[x] = \sin[x]^2$ and $g[x] = x^2 \cos[x]^{2/3}$ on the same axes for $-1 \le x \le 1$.

2 Lesson 4.01 Splines

```
In[2]:=
  Clear[x,f,g]
  f[x_] = Sin[x]^2;
  g[x_] = x^2 Cos[x]^(2/3);
  Plot[{f[x],g[x]},{x,-1,1},AxesLabel->{"x",""},
  PlotStyle->{{GrayLevel[0.4],Thickness[0.01]},
  {Red,Thickness[0.01]}}];
```

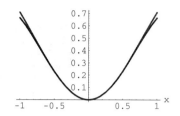

Describe what you see.

Answer: Very similar plots near $x = 0$. As you advance x from the left of 0 to the right of 0, you can make the transition from one curve to the other with ease.

B.1.a.iii) Here are plots of
$$f[x] = 1 + \sin[x]$$
and
$$g[x] = \frac{60 + 60\,x + 3\,x^2 - 7\,x^3}{60 + 3\,x^2}$$
on the same axes for $-3 \le x \le 3$.

```
In[3]:=
  f[x_] = 1 + Sin[x];
  g[x_] = (60 + 60 x + 3 x^2 - 7 x^3)/(60 + 3 x^2);
  Plot[{f[x],g[x]},{x,-3,3},AxesLabel->{"x",""},
  PlotStyle->{{GrayLevel[0.4],Thickness[0.01]},
  {Red,Thickness[0.01]}}];
```

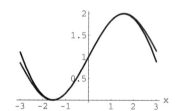

Describe what you see.

Answer: Very similar plots near $x = 0$. As you advance x from the left of 0 to the right of 0, you can make the transition from one curve to the other with ease.

B.1.a.iv) Here are plots of
$$f[x] = \cos[x],$$
and
$$g[x] = -1 + \left(\frac{\pi^2}{2}\right)(x-1)^2$$
on the same axes for $0 \le x \le 2$:

In[4]:=
```
Clear[x,f,g]
f[x_] = Cos[Pi x];
g[x_] = -1 + (Pi^2/2) (x - 1)^2;
Plot[{f[x],g[x]},{x,0,2},AxesLabel->{"x",""},
PlotStyle->{{GrayLevel[0.4],Thickness[0.01]},
{Red,Thickness[0.01]}}];
```

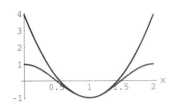

Describe what you see.

Answer: Very similar plots near $x = 1$. As you advance x from the left of 1 to the right of 1, you can make the transition from one curve to the other with ease.

B.1.b.i) Are the plotting phenomena you saw in the last four plots accidents?

Answer: Get real. In mathematics, there are no accidents.

B.1.b.ii) What can explain what's happening?

Answer: It has to do with derivatives. Look at the first plot again:

In[5]:=
```
Clear[x,f,g]
f[x_] = Cos[2 x];
g[x_] = Sqrt[1 - 4 x^2];
Plot[{f[x],g[x]},{x,-0.5,0.5},
AxesLabel->{"x",""},PlotStyle->
{{GrayLevel[0.4],Thickness[0.01]},
{Red,Thickness[0.01]}}];
```

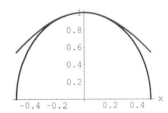

Very similar plots near $x = 0$. As you advance x from the left of 0 to the right of 0, you can make the transition from one curve to the other with ease. Now look at:

In[6]:=
```
{f[0],g[0]}
```
Out[6]=
```
{1, 1}
```

In[7]:=
```
{f'[0],g'[0]}
```
Out[7]=
```
{0, 0}
```

In[8]:=
```
{f''[0],g''[0]}
```

Out[8]=
{-4, -4}

In[9]:=
{f'''[0],g'''[0]}

Out[9]=
{0, 0}

In[10]:=
{f''''[0],g''''[0]}

Out[10]=
{16, -48}

Both functions go through $\{0,1\}$ and they have the same first, second, and third derivatives at $x = 0$. Now look at the plot from B.1.a.ii):

In[11]:=
```
Clear[x,f,g]
f[x_] = Sin[x]^2;
g[x_] = x^2 Cos[x]^(2/3);
Plot[{f[x],g[x]},{x,-1,1},AxesLabel->{"x",""},
PlotStyle->{{GrayLevel[0.4],Thickness[0.01]},
{Red,Thickness[0.01]}}];
```

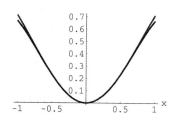

Very similar plots near $x = 0$. As you advance x from the left of 0 to the right of 0, you can make the transition from one curve to the other with ease.

Look at:

In[12]:=
{f[0],g[0]}

Out[12]=
{0, 0}

Both functions go through $\{0,0\}$. Look at a table of the first eight derivatives, all calculated at $x = 0$:

In[13]:=
```
Clear[k]
Table[{D[f[x],{x,k}],D[g[x],{x,k}]}/.x->0,{k,1,8}]
```

Out[13]=
{{0, 0}, {2, 2}, {0, 0}, {-8, -8}, {0, 0}, {32, 0}, {0, 0}, {-128, -($\frac{896}{9}$)}}

Both functions go through $\{0,0\}$ and they have the same first, second, third, fourth, and fifth derivatives at $x = 0$.

B.1.b.iii) The upshot of all this: Agree that $f[x]$ and $g[x]$ have order of contact m at $x = a$ if
$$f[a] = g[a], f'[a] = g'[a], f''[a] = g''[a], \ldots, f^{(m-1)}[a] = g^{(m-1)}[a]$$
and
$$f^{(m)}[a] = g^{(m)}[a],$$
so that the functions and their first m derivatives agree at $x = a$.

The message of the plots above is this: The higher the order of contact of $f[x]$ and $g[x]$ at $x = a$, then the smoother the transition from one curve to the other as you advance x from the left of a to the right of a.

> How does the Race Track Principle begin to explain this fact?

Answer: Here's a version of the Race Track Principle which explains the smooth transition in the case of order of contact 1:

Horses If two horses are tied at one point, and they run at the same speed at this point, then they run close together near this point.

Functions If $f[a] = g[a]$ and $f'[a] = g'[a]$, then the two functions plot out nearly the same as x advances from the a little bit to the left of $x = a$ to a little bit to the right of $x = a$. This explains why when you have order of contact 1 at $x = a$, then you have a smooth transition as you advance x from the left of $x = a$ to the right of $x = a$.

To see why it is that when you have order of contact 2 at $x = a$, then you can expect an even smoother transition as you advance x from the left of $x = a$ to the right of $x = a$, remember that order of contact 2 at $x = a$ means

$$f[a] = g[a],$$
$$f'[a] = g'[a]$$

and

$$f''[a] = g''[a].$$

The fact that $f''[a] = g''[a]$ and $f'[a] = g'[a]$ ensures that $f'[x]$ is very close to $g'[x]$ as x advances from the left of $x = a$ to the right of $x = a$. This means that $f[x]$ is very, very close to $g[x]$ as x advances from the left of $x = a$ to the right of $x = a$. To see the difference, look at these:

In[14]:=
```
Clear[x,f,g]; f[x_] = 1/(1 - x); g[x_] = 1 + x;
Table[{D[f[x],{x,k}],D[g[x],{x,k}]}/.x->0,{k,0,2}]
```

Out[14]=
```
{{1, 1}, {1, 1}, {2, 0}}
```

The order of contact is 1 at $x = 0$. Here are plots:

In[15]:=
```
Plot[{f[x],g[x]},{x,-0.5,0.5},
 PlotStyle->{{GrayLevel[0.4],Thickness[0.01]},
 {Red,Thickness[0.01]}},AxesLabel->{"x",""},
 PlotRange->{0.5,2},PlotLabel->
 "Order of contact 1 at x = 0"];
```

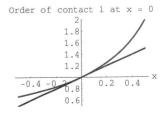

Now look at these:

In[16]:=
```
Clear[x,f,g]; f[x_] = 1/(1 - x); g[x_] = 1 + x + x^2;
Table[{D[f[x],{x,k}],D[g[x],{x,k}]}/.x->0,{k,0,3}]
```
Out[16]=
{{1, 1}, {1, 1}, {2, 2}, {6, 0}}

The order of contact is 2 at $x = 0$. Here are plots:

In[17]:=
```
Plot[{f[x],g[x]},{x,-0.5,0.5},
PlotStyle->{{GrayLevel[0.4],Thickness[0.01]},
{Red,Thickness[0.01]}},AxesLabel->{"x",""},
PlotRange->{0.5,2},PlotLabel->
"Order of contact 2 at x = 0"];
```

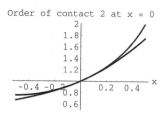

The higher the order of contact, the smoother the transition.

■ B.2) Smooth splines

A spline is a long flexible strip of plastic or the like that is used in drawing smooth curves. Derivatives can be used for the same purpose. Suppose you have two functions $f[x]$ and $g[x]$, and you have a number b. Create a new function, spline$[x]$, by setting

$$\text{spline}[x] = f[x] \quad \text{for } x \leq b$$

and

$$\text{spline}[x] = g[x] \quad \text{for } x > b.$$

If $f[b] = g[b]$, then most folks say that the new function spline$[x]$ is a spline of $f[x]$ and $g[x]$ knotted at $\{b, f[b]\} = \{b, g[b]\}$.

B.2.a) Here's a spline of

$$f[x] = e^x \quad \text{and} \quad g[x] = \sqrt{\frac{1+x}{1-x}} + \sin[6\,x^3]$$

knotted at $\{0, 1\}$ which is plotted on an interval including 0.

In[18]:=
```
Clear[f,g,spline,x]; f[x_] = E^x;
g[x_] = Sqrt[(1 + x)/(1 - x)] + Sin[6 x^3];
spline[x_] := g[x]/;x > 0;
spline[x_] := f[x]/;x <= 0;
splineplot = Plot[spline[x],{x,-0.8,0.8},
PlotStyle->{{Blue,Thickness[0.02]}},
AxesLabel->{"x",""},PlotLabel->
"spline knotted at {0,1}"];
```

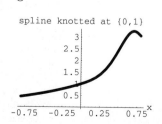

> Discuss reasons for the smoothness of the spline curve as it passes through the knot at $\{0, 1\}$.

Answer: To see why spline$[x]$ is so smooth at the knot at $\{0, 1\}$, look at:

In[19]:=
```
{{f[0],g[0]}, {f'[0],g'[0]}, {f''[0],g''[0]}, {f'''[0],g'''[0]}}
```

Out[19]=
```
{{1, 1}, {1, 1}, {1, 1}, {1, 39}}
```

The smoothness of the spline at the knot at $\{0, 1\}$ comes from the fact that $f[x]$ and $g[x]$ have order of contact 2 at the knot at $\{0, 1\}$. Just to confirm, look at a plot of $f[x]$ and $g[x]$ together.

In[20]:=
```
fgplot = Plot[{f[x],g[x]},{x,-0.8,0.8},
  PlotStyle->{{Red},{GrayLevel[0.3]}},
  AxesLabel->{"x",""}];
```

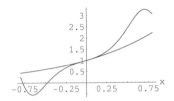

Yep. The fact that $f[x]$ and $g[x]$ have order of contact 2 makes for a smooth transition from $f[x]$ to $g[x]$ at $x = 0$. This smooth transition is what's responsible for the smoothness in the spline at $\{0, 1\}$. Now look at a plot of $f[x], g[x]$, and the spline $h[x]$:

In[21]:=
```
Show[splineplot,fgplot,PlotLabel->None];
```

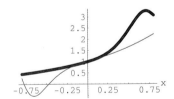

Nice transition.

B.2.b) Here's a spline of

$$f[x] = \frac{1}{1+x^2} \quad \text{and} \quad g[x] = \frac{9}{8} - \frac{x}{2} - \frac{x^2}{2} + \frac{x^3}{2} - \frac{x^4}{8}$$

knotted at $\{1, 0.5\}$ and plotted on an interval including 1.

In[22]:=
```
Clear[f,g,spline,x]; f[x_] = 1/(1 + x^2);
g[x_] = 9/8 - x/2 - x^2/2 + x^3/2 - x^4/8;
spline[x_] := g[x]/;x > 1 ;
spline[x_] := f[x]/;x <= 1;
splineplot = Plot[spline[x],{x,0,2},
  PlotStyle->{{Blue,Thickness[0.01]}},
  PlotRange->{0,1.3},AxesLabel->{"x",""},
  PlotLabel->"spline knotted at {1,0.5}"];
```

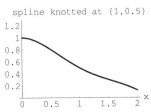

8 Lesson 4.01 Splines

> Discuss reasons for the smoothness of the spline curve as it passes through the knot at $\{1, 0.5\}$.

Answer: Look at:

In[23]:=
```
{{f[1],g[1]}, {f'[1],g'[1]}, {f''[1],g''[1]}, {f'''[1],g'''[1]},
 {f''''[1],g''''[1]}, {f'''''[1],g'''''[1]}}
```

Out[23]=
$$\{\{\tfrac{1}{2}, \tfrac{1}{2}\}, \{-(\tfrac{1}{2}), -(\tfrac{1}{2})\}, \{\tfrac{1}{2}, \tfrac{1}{2}\}, \{0, 0\}, \{-3, -3\}, \{15, 0\}\}$$

The smoothness of the spline at the knot at $\{1, 0.5\}$ comes from the fact that $f[x]$ and $g[x]$ have order of contact 4 at $x = 1$. Just to confirm, here's a plot of $f[x]$ and $g[x]$:

In[24]:=
```
fgplot = Plot[{f[x],g[x]},{x,0,2},
  PlotStyle->{{Red},{GrayLevel[0.2]}},
  PlotRange->{0,1.3},AxesLabel->{"x",""}];
```

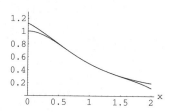

Yep. A beautiful transition at $\{1, 0.5\}$.

Now see a plot of $f[x]$, $g[x]$, and the spline $h[x]$:

In[25]:=
```
Show[splineplot,fgplot,PlotLabel->None];
```

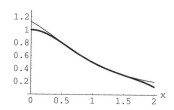

Nice transition.

B.2.c) Here's a spline of x^2 and $-x$ knotted at $\{0, 0\}$, plotted on an interval including 0.

In[26]:=
```
Clear[f,g,spline,x]; f[x_] = x^2; g[x_] = -x;
spline[x_] := g[x]/; x > 0;
spline[x_] := f[x]/; x <= 0;
splineplot = Plot[spline[x],{x,-1,1},
  PlotStyle->{{Blue,Thickness[0.01]}},
  PlotRange->{-1,1},AxesLabel->{"x",""},
  PlotLabel->"spline knotted at {1,0.0}"];
```

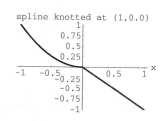

> Discuss reasons for the lack of smoothness of the spline curve as it passes through the knot at $\{0,0\}$.

Answer: Look at:

In[27]:=
```
{{f[0],g[0]}, {f'[0],g'[0]}}
```

Out[27]=
```
{{0, 0}, {0, -1}}
```

The order of contact at $x = 0$ is 0. This explains the lack of smoothness of the spline at the knot at $\{0,0\}$. Just to confirm, look at a plot of $f[x]$ and $g[x]$.

In[28]:=
```
fgplot = Plot[{f[x],g[x]},{x,-1,1},
  PlotStyle->{{Red},{GrayLevel[0.2]}},
  PlotRange->{-1,1},AxesLabel->{"x",""}];
```

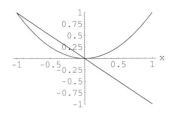

Terrible transition at $\{0,0\}$. Now see a plot of $f[x], g[x]$ and $spline[x]$:

In[29]:=
```
Show[splineplot,fgplot,PlotLabel->None];
```

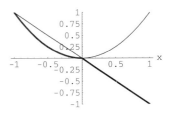

This type of transition would make a very dangerous interstate highway connection.

B.2.d) > What practical good are smooth splines?

Answer: Good question. First of all, the idea of a spline smooth at its knots makes for interesting plots and nice art. But there's a serious aspect as well: Laying out the curves on expressways and railroads amounts to joining curved plots to relatively straight plots at knots of a spline. It is unacceptable to have a hitch (corner) in the middle of the roadway. The higher the order of contact at the knots, the safer the highway or roadbed. Any order of contact less than 2 is considered unsafe, and this may be pushing it.

Tutorials

■ T.1) Splining functions and polynomials

T.1.a.i) Find the polynomial $g[x]$ of degree 2 that has the highest possible order of contact with $f[x] = e^x$ at $x = 0$.

Plot the spline knotted at $\{0, 1\}$ with e^x on the right and your polynomial on the left. Also show a plot of $f[x]$, the polynomial, and the spline.

Answer: Enter $f[x]$ and the general form of a second degree polynomial:

In[1]:=
```
Clear[f,g,x,a,k]; f[x_] = E^x;
g[x_] = Sum[a[k] x^k,{k,0,2}]
```
Out[1]=
$$a[0] + x\, a[1] + x^2\, a[2]$$

You need:

In[2]:=
```
eqn1 = f[0] == g[0]
eqn2 = f'[0] == g'[0]
eqn3 = f''[0] == g''[0]
```
Out[2]=
```
1 == a[0]
1 == a[1]
1 == 2 a[2]
```

Now you've got three equations in the three unknowns $a[1], a[2]$ and $a[3]$. Going for a fourth equation will only overdetermine the system of equations, so solve:

In[3]:=
```
coeffs = Solve[{eqn1,eqn2,eqn3}]
```
Out[3]=
$$\{\{a[0] \to 1,\ a[1] \to 1,\ a[2] \to \tfrac{1}{2}\}\}$$

The polynomial you are after is:

In[4]:=
```
Clear[poly]; poly[x_] = g[x]/.coeffs[[1]]
```
Out[4]=
$$1 + x + \frac{x^2}{2}$$

Check:

In[5]:=
```
Table[{D[f[x],{x,k}],D[poly[x],{x,k}]}/.x->0,{k,0,3}]
```
Out[5]=
```
{{1, 1}, {1, 1}, {1, 1}, {1, 0}}
```

This gives you order of contact 2 between $f[x]$ and the polynomial at $x = 0$. And you cannot do better. Here comes the plot of the spline knotted at $\{0, 1\}$ with e^x on the right and the polynomial on the left.

In[6]:=
```
Clear[spline]
spline[x_] := poly[x]/;x <= 0;
spline[x_] := f[x]/;0 < x;
splineplot = Plot[spline[x],{x,-2,2},
PlotStyle->{{Blue,Thickness[0.01]}},
AxesLabel->{"x",""},PlotLabel->
"spline knotted at {0,1}"];
```

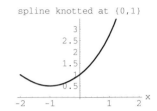

Smooth. Here's the plot of $f[x]$, the polynomial, and the spline:

In[7]:=
```
Plot[{spline[x],f[x],poly[x]},{x,-2,2},
PlotStyle->{{Blue,Thickness[0.02]},
{Red,Thickness[0.01]},
{Red,Thickness[0.01]}},
AxesLabel->{"x",""}];
```

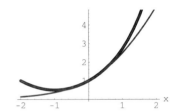

Beautiful transition.

T.1.a.ii) Find the polynomial $g[x]$ of degree 6 that has the highest possible order of contact with $f[x] = e^x$ at $x = 0$.

Plot the spline knotted at $\{0, 1\}$ with e^x on the right and your polynomial on the left. Also show a plot of $f[x]$, the polynomial, and the spline.

Answer: Enter $f[x]$ and the general form of a sixth degree polynomial:

In[8]:=
```
Clear[f,g,x,a,k]; f[x_] = E^x;
g[x_] = Sum[a[k] x^k,{k,0,6}]
```

Out[8]=
$$a[0] + x\, a[1] + x^2\, a[2] + x^3\, a[3] + x^4\, a[4] + x^5\, a[5] + x^6\, a[6]$$

This leaves seven unknowns $a[0]$, $a[1]$, $a[2]$, $a[3]$, $a[4]$, $a[5]$, and $a[6]$ to determine. You need seven equations to do this:

In[9]:=
```
eqn1 = f[0] == g[0]
eqn2 = f'[0] == g'[0]
eqn3 = f''[0] == g''[0]
eqn4 = f'''[0] == g'''[0]
eqn5 = f''''[0] == g''''[0]
eqn6 = f'''''[0] == g'''''[0]
eqn7 = f''''''[0] == g''''''[0]
```

Out[9]=
```
1 == a[0]
1 == a[1]
1 == 2 a[2]
1 == 6 a[3]
1 == 24 a[4]
1 == 120 a[5]
1 == 720 a[6]
```

Now you've got the seven equations in the seven unknowns. Going for an eighth equation will only overdetermine the system of equations, so solve:

In[10]:=
```
coeffs = Solve[{eqn1,eqn2,eqn3,eqn4,eqn5,eqn6,eqn7}]
```

Out[10]=

$\{\{a[0] \to 1,\ a[1] \to 1,\ a[2] \to \frac{1}{2},\ a[3] \to \frac{1}{6},\ a[4] \to \frac{1}{24},\ a[5] \to \frac{1}{120},\ a[6] \to \frac{1}{720}\}\}$

The polynomial you are after is:

In[11]:=
```
Clear[poly]; poly[x_] = g[x]/.coeffs[[1]]
```

Out[11]=

$1 + x + \frac{x^2}{2} + \frac{x^3}{6} + \frac{x^4}{24} + \frac{x^5}{120} + \frac{x^6}{720}$

Check:

In[12]:=
```
Table[{D[f[x],{x,k}],D[poly[x],{x,k}]}/.x->0,{k,0,7}]
```

Out[12]=
```
{{1, 1}, {1, 1}, {1, 1}, {1, 1}, {1, 1}, {1, 1}, {1, 1}, {1, 0}}
```

This gives you order of contact 6 between $f[x]$ and the polynomial at $x = 0$. And you cannot do better.

Here comes the plot of the spline knotted at $\{0,1\}$ with e^x on the right and the polynomial on the left.

In[13]:=
```
Clear[spline]
spline[x_] := poly[x]/; x <= 0;
spline[x_] := f[x]/; 0 < x;
splineplot = Plot[spline[x],{x,-2,2},
PlotStyle->{{Blue,Thickness[0.01]}},
AxesLabel->{"x",""},PlotLabel->
"spline knotted at {0,1}"];
```

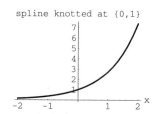

Very, very smooth. Here's the plot of $f[x]$, the polynomial, and the spline:

In[14]:=
```
Plot[{spline[x],
f[x],poly[x]},{x,-2,2},
PlotStyle->{{Blue,Thickness[0.02]},
{Thickness[0.01]},{Red,Thickness[0.01]}},
AxesLabel->{"x",""}];
```

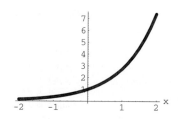

Such a beautiful transition that you can't tell the functions apart without a scorecard. See them on a longer interval centered at 0:

In[15]:=
```
Plot[{spline[x],
f[x],poly[x]},{x,-4,4},
PlotStyle->{{Blue,Thickness[0.02]},
{Thickness[0.01]},{Red,Thickness[0.01]}},
AxesLabel->{"x",""}];
```

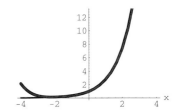

A very long, very smooth transition.

■ T.2) Landing an airplane

The design of an electronically controlled airplane landing system calls for the plane to approach the runway head-on, at a constant horizontal speed and constant altitude h. As the plane passes over a certain point on the ground R feet from the designated touch-down spot on the runway, the system is to take over and bring the plane onto the runway on the trajectory of a polynomial. The constant horizontal speed d is to be maintained throughout the whole landing procedure.

Agree that altitude[x] stands for the altitude of the plane when the plane is directly above a spot on the ground x feet from the designated touch-down spot on the runway.

To achieve smoothness at the beginning of the descent at R feet from the designated touch-down spot on the runway and at the landing on the runway, you need a smooth spline between horizontal lines $y = h$ and $y = 0$. To go for order of contact 2 at the knots at $\{0, 0\}$ and $\{R, h\}$, you need:

$$\text{altitude}[0] = 0,$$
$$\text{altitude}'[0] = 0,$$
$$\text{altitude}''[0] = 0,$$
$$\text{altitude}[R] = h,$$
$$\text{altitude}'[R] = 0,$$

and

$$\text{altitude}''[R] = 0.$$

These six equations give you six degrees of freedom. A fifth degree polynomial has six unspecified constants, so try:

In[16]:=
```
Clear[a,altitude,x,h,R]
altitude[x_] = Sum[a[k] x^k,{k,0,5}]
```

Out[16]=
$$a[0] + x\, a[1] + x^2\, a[2] + x^3\, a[3] + x^4\, a[4] + x^5\, a[5]$$

Now feed in the equations and solve:

In[17]:=
```
eqn1 = altitude[0] == 0;
eqn2 = altitude'[0] == 0;
eqn3 = altitude''[0] == 0;
eqn4 = altitude[R] == h;
eqn5 = altitude'[R] == 0;
eqn6 = altitude''[R] == 0;
solutions = Solve[{eqn1,eqn2,eqn3,eqn4,eqn5,eqn6},
  {a[0],a[1],a[2],a[3],a[4],a[5]}]
```

Out[17]=

$$\{\{a[0] \to 0, a[1] \to 0, a[2] \to 0, a[3] \to \frac{10\,h}{R^3}, a[4] \to \frac{-15\,h}{R^4}, a[5] \to \frac{6\,h}{R^5}\}\}$$

Here is the function that describes the landing:

In[18]:=
```
Clear[trajectory]
trajectory[x_,h_,R_] = altitude[x]/.solutions[[1]]
```

Out[18]=

$$\frac{10\,h\,x^3}{R^3} - \frac{15\,h\,x^4}{R^4} + \frac{6\,h\,x^5}{R^5}$$

T.2.a) Plot the automatic landing trajectory for the case $R = 10000$ ft and $h = 5000$ ft.

Answer: The plot of the landing trajectory is:

In[19]:=
```
R = 10000; h = 5000;
landing = Plot[trajectory[x,h,R],{x,0,R},
  PlotStyle->{{Blue,Thickness[0.01]}},
  AxesLabel->{"x","altitude"},
  PlotLabel->"Actual Landing"];
```

Here's a plot showing the plane's path slightly before the descent begins, and slightly after the plane touches down:

In[20]:=
```
Clear[spline]
spline[x_] := 0/; x <= 0;
spline[x_] := trajectory[x,h,R]/; 0 < x < R;
spline[x_] := h/; R <= x;
bigpicture = Plot[spline[x],{x,-1000,R + 2000},
  PlotStyle->{{Blue,Thickness[0.01]}},
  AxesLabel->{"x","altitude"},
  PlotLabel->"Big picture"];
```

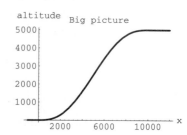

How beautiful to fly. See how the smoothness in the spline at the knots at $\{0,0\}$ and $\{R,h\}$ guarantees a smooth landing.

■ T.3) Does *Mathematica*'s Interpolation function give you a smooth spline?

T.3.a) The default option in *Mathematica*'s Interpolation function attempts to fit given data with consecutive third degree polynomials. The question here: Is *Mathematica* programmed to make the interpolation function a smooth spline as it shifts from one polynomial to the next?

Answer: Try it on some actual data and see:

Here's a plot of United States national debt data in the form $\{x,y\}$, where x is a year after 1980, and y is the national debt for year x.

```
In[21]:=
  nationaldebt = {{1980,907.7},{1981,997.9},
  {1982,1142},{1983,1377.2},{1984,1572.3},
  {1985,1823.1},{1986,2125.3},{1987,2530.3},
  {1988,2602.3},{1989,2857.4},{1990,3233.3}};
  dataplot = ListPlot[nationaldebt,
  AxesLabel->{"x","y"},
  PlotStyle->{Red,PointSize[0.02]}];
```

Here is *Mathematica*'s Interpolation function through these data and a plot:

```
In[22]:=
  Clear[debt]
  debt[t_] = Interpolation[nationaldebt][t];
  functplot = Plot[debt[t],{t,1980,1990},
  PlotStyle->{{Thickness[0.01],GrayLevel[0.5]}},
  PlotRange->All,DisplayFunction->Identity];
  Show[functplot,dataplot,AxesLabel->
  {"t","National Debt\n in millions"},
  DisplayFunction->$DisplayFunction];
```

It looks close to being smooth, but there seem to be some little corners between 1981 and 1985. Probably it is not a smooth spline. One way to check for sure is to plot its derivative:

```
In[23]:=
  Plot[debt'[t],{t,1981,1984},
  PlotStyle->{{Thickness[0.01],Red}},
  PlotRange->All];
```

Wild jumps in the derivative in 1982 and 1983. This is a dead giveaway that *Mathematica*'s Interpolation function is not a smooth spline of third degree polynomials. If you want to see the labor involved in making a smooth spline of third degree polynomials, see the Give It a Try section.

Give It a Try

Experience with the starred ($\star$) problems will be especially beneficial for understanding later lessons.

■ G.1) Explain the plots*

G.1.a) Here's a spline of

$$f[x] = x^2 \sin[3\,x] \quad \text{and} \quad g[x] = x^2 \sin[6\,x]$$

knotted at $\{0, 0\}$ abd plotted on an interval including 0.

In[1]:=
```
Clear[f,g,spline,x]
f[x_] = x^2 Sin[3 x]; g[x_] = x^2 Sin[6 x];
spline[x_] := g[x]/; x > 0;
spline[x_] := f[x]/; x <= 0;
splineplot = Plot[spline[x],{x,-3,3},
PlotStyle->{{Blue,Thickness[0.02]}},
AxesLabel->{"x",""},PlotLabel->
"spline knotted at {0,0}"];
```

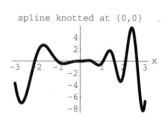

Discuss reasons for the smoothness (or lack of smoothness) of the spline curve as it passes through the knot at $\{0, 0\}$.

G.1.b.i) Here's a spline of

$$f[x] = \sin[\pi\,x] \quad \text{and} \quad (x-1)^2 \cos[x]$$

knotted at $\{1, 0\}$ and plotted on an interval including 1.

In[2]:=
```
Clear[f,g,spline,x]
f[x_] = Sin[Pi x]; g[x_] = (x - 1)^2 Cos[x];
spline[x_] := g[x]/; x > 1;
spline[x_] := f[x]/; x <= 1;
splineplot = Plot[spline[x],{x,0,2},
PlotStyle->{{Blue,Thickness[0.02]}},
AxesLabel->{"x",""},PlotLabel->
"spline knotted at {1,0}"];
```

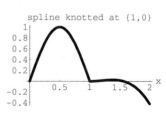

Discuss reasons for the smoothness (or lack of smoothness) of the spline curve as it passes through the knot at $\{1, 0\}$.

G.1.b.ii) Here's a spline of
$$f[x] = e^{-x^2} \quad \text{and} \quad g[x] = \cos[\sqrt{2}\,x]$$
knotted at $\{0, 1\}$ and plotted on an interval including 0.

In[3]:=
```
Clear[f,g,spline,x]
f[x_] = E^(-x^2);
g[x_] = Cos[Sqrt[2] x];
spline[x_] := g[x]/; x > 0;
spline[x_] := f[x]/; x <= 0;
splineplot = Plot[spline[x],{x,-1,1},
PlotStyle->{{Blue,Thickness[0.02]}},
AxesLabel->{"x",""},PlotLabel->
"spline knotted at {0,1}"];
```

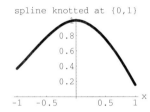

Discuss reasons for the smoothness (or lack of smoothness) of the spline curve as it passes through the knot at $\{0, 0\}$.

G.1.c) Here are plots of
$$f[x] = \arcsin[x] \quad \text{and} \quad g[x] = \frac{3x}{2 + \sqrt{1 - x^2}}$$
on the same axes for $-1 \leq x \leq 1$:

In[4]:=
```
Clear[f,g,x];
f[x_] = ArcSin[x];
g[x_] = 3 x/(2 + Sqrt[1 - x^2]);
Plot[{f[x],g[x]},{x,-1,1},
AxesLabel->{"x",""},PlotStyle->
{{SkyBlue,Thickness[0.02]},
{IndianRed,Thickness[0.01]}}];
```

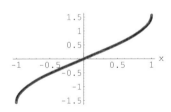

Many calculus students in the old-fashioned calculus course would be hard pressed to explain why the plots turned out the way they did. But you can use order of contact at $x = 0$ to say why you think they came out this way.

Do it.

G.1.d) Determine constants a, b, and c so that
$$f[x] = \sin[x] + \sin[x^2] \quad \text{and} \quad g[x] = a + b e^{cx}$$
make a smooth spline knotted at $\{0, 0\}$. Plot to confirm.

■ G.2) Splining functions and polynomials*

G.2.a.i) Find the polynomial of degree 3 that has the highest possible order of contact with $f[x] = \sin[x]$ at $x = 0$.

Plot the spline knotted at $\{0, 0\}$ with $f[x]$ on the right and your polynomial on the left for $-2 \leq x \leq 2$.

Also plot on the same axes $f[x]$, the polynomial, and the spline for $-2 \leq x \leq 2$. Describe what you see.

G.2.a.ii) Find the polynomial of degree 5 that has the highest possible order of contact with $f[x] = \sin[x]$ at $x = 0$.

Plot the spline knotted at $\{0, 0\}$ with $f[x]$ on the right and your polynomial on the left for $-2 \leq x \leq 2$.

Also plot on the same axes $f[x]$, the polynomial, and the spline for $-2 \leq x \leq 2$.

Describe what you see, paying special attention to a comparison of the plots in parts G.2.a.i) and G.2.a.ii).

G.2.b.i) Find the polynomial of degree 2 that has the highest possible order of contact with $f[x] = \cos[x]$ at $x = 0$.

Plot the spline knotted at $\{0, 1\}$ with $f[x]$ on the right and your polynomial on the left for $-2 \leq x \leq 2$.

Also plot on the same axes $f[x]$, the polynomial, and the spline for $-2 \leq x \leq 2$. Describe what you see.

G.2.b.ii) Find the polynomial of degree 4 that has the highest possible order of contact with $f[x] = \cos[x]$ at $x = 0$.

Plot the spline knotted at $\{0, 1\}$ with $f[x]$ on the right and your polynomial on the left for $-2 \leq x \leq 2$.

Also plot on the same axes $f[x]$, the polynomial, and the spline for $-2 \le x \le 2$.

Describe what you see, paying special attention to a comparison of the plots in parts B.2.b.i) and B.2.b.ii).

G.2.c.i) Find the polynomial of degree 2 that has the highest possible order of contact with $f[x] = 1/(1-x)$ at $x = 0$.

Plot the spline knotted at $\{0, 1\}$ with $f[x]$ on the right and your polynomial on the left for $-0.9 \le x \le 0.9$.

Also plot on the same axes $f[x]$, the polynomial, and the spline for $-0.9 \le x \le 0.9$. Describe what you see.

G.2.c.ii) Find the polynomial of degree 4 that has the highest possible order of contact with $f[x] = 1/(1-x)$ at $x = 0$.

Plot the spline knotted at $\{0, 1\}$ with $f[x]$ on the right and your polynomial on the left for $-0.9 \le x \le 0.9$.

Also plot on the same axes $f[x]$, the polynomial, and the spline for $-0.9 \le x \le 0.9$.

Describe what you see, paying special attention to a comparison of the plots in parts G.2.c.i) and G.2.c.ii).

G.2.d.i) Find the polynomial of degree 1 that has the highest possible order of contact with $f[x] = \text{Erf}[x-1]$ at $x = 1$.

Plot the spline knotted at $\{1, 0\}$ with $f[x]$ on the right and your polynomial on the left for $0 \le x \le 2$.

Also plot on the same axes $f[x]$, the polynomial, and the spline for $0 \le x \le 2$. Describe what you see.

G.2.d.ii) Find the polynomial of degree 3 that has the highest possible order of contact with $f[x] = \text{Erf}[x-1]$ at $x = 1$.

Plot the spline knotted at $\{1, 0\}$ with $f[x]$ on the right and your polynomial on the left for $0 \le x \le 2$.

Also plot on the same axes $f[x]$, the polynomial, and the spline for $0 \le x \le 2$.

Describe what you see, paying special attention to a comparison of the plots in parts G.2.d.i) and G.2.d.ii).

■ G.3) Splines in road design

G.3.a) Here are two straight roads running parallel to each other:

```
In[5]:=
  Clear[high,low,x];
  high[x_] = 1; low[x_] = - 1;
  roads = Plot[{high[x],low[x]},{x,-4,4},
  PlotStyle->{{DarkSlateGray,Thickness[0.02]},
  {DarkSlateGray,Thickness[0.02]}},
  AxesLabel->{"x",""},PlotRange->{-2,2}];
```

> Design a smooth crossover that leaves the bottom road at $\{-3,-1\}$ and connects to the top road at $\{3,1\}$. For safety, make sure your crossover has at least order of contact 2 at both connections. Plot your crossover.

G.3.b) Here are two curved roads:

```
In[6]:=
  Clear[high,low,x]
  high[x_] = (x^2)/4 + 1;
  low[x_] = -(x^2)/4 - 1;
  roads = Plot[{high[x],low[x]}, {x,-4,4},
  PlotStyle->{{DarkSlateGray,Thickness[0.02]},
  {DarkSlateGray,Thickness[0.02]}},
  AxesLabel->{"x",""}];
```

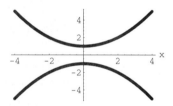

> Design a smooth crossover that leaves the bottom road at $\{-2,-2\}$, and connects to the top road at $\{1,5/4\}$. For safety, make sure your crossover has at least order of contact 2 at both connections. Plot your crossover.

■ G.4) Order of contact for derivatives and integrals*

G.4.a.i) Take $f[x] = x \sin[x^2]$ and $g[x] = x \sin^2[x]$ and look at:

```
In[7]:=
  Clear[x,f,g,k]; f[x_] = x Sin[x^2]; g[x_] = x^2 Sin[x];
  Table[{D[f[x],{x,k}],D[g[x],{x,k}]}/.x->0,{k,0,5}]

Out[7]=
  {{0, 0}, {0, 0}, {0, 0}, {6, 6}, {0, 0}, {0, -20}}
```

$f[x]$ and $g[x]$ have order of contact 4 at $x = 0$. Now put

$$F[x] = f'[x] \quad \text{and} \quad G[x] = g'[x]:$$

> What is the order of contact of $F[x]$ and $G[x]$ at $x = 0$?

G.4.a.ii) Take any two functions $f[x]$ and $g[x]$ that have order of contact 6 at $x = b$. Put $F[x] = f'[x]$ and $G[x] = g'[x]$.

> What do you think the order of contact of $F[x]$ and $G[x]$ is at $x = b$?

G.4.b.i) Take $f[x] = x\, e^x$ and $g[x] = \sin[x]\, e^x$ and look at:

In[8]:=
```
Clear[x,f,g,k]; f[x_] = x E^x; g[x_] = Sin[x] E^x;
Table[{D[f[x],{x,k}],D[g[x],{x,k}]}/.x->0,{k,0,5}]
```

Out[8]=
{{0, 0}, {1, 1}, {2, 2}, {3, 2}, {4, 0}, {5, -4}}

$f[x]$ and $g[x]$ have order of contact 2 at $x = 0$. Now put $F[x] = \int_0^x f[t]\, dt$ and $G[x] = \int_0^x g[t]\, dt$.

> What is the order of contact of $F[x]$ and $G[x]$ at $x = 0$?

G.4.b.ii) Take any two functions $f[x]$ and $g[x]$ that have order of contact 3 at $x = b$. Put $F[x] = \int_b^x f[t]\, dt$ and $G[x] = \int_b^x g[t]\, dt$.

> What do you think the order of contact of $F[x]$ and $G[x]$ is at $x = b$?

G.4.c) Here are two functions whose plots cross at $x = 0$.

In[9]:=
```
Clear[f,g,x]
f[x_] = E^x; g[x_] = 1 - x;
Plot[{f[x],g[x]},{x,-1,1},
 PlotStyle->{{Thickness[0.01]},
 {Red,Thickness[0.01]}},
 AxesLabel->{"x",""}];
```

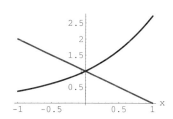

As you can see, there is no hope of making a smooth spline knotted at $\{0, 1\}$ from these two functions; so they have order of contact 0 at $x = 0$. Check:

In[10]:=
```
{{f[0],g[0]}, {f'[0],g'[0]}}
```

Out[10]=
{{1, 1}, {1, -1}}

Yep; their order of contact is 0 at $x = 0$.

Now put $F[x] = \int_0^x f[t]\, dt$ and $G[x] = \int_0^x g[t]\, dt$ and plot:

In[11]:=
```
Clear[F,G,t]; F[x_] = Integrate[f[t],{t,0,x}]
```
Out[11]=
$$-1 + E^x$$

In[12]:=
```
G[x_] = Integrate[g[t],{t,0,x}]
```
Out[12]=
$$x - \frac{x^2}{2}$$

In[13]:=
```
Plot[{F[x],G[x]},{x,-1,1},
 PlotStyle->{{Thickness[0.01]},
 {Red,Thickness[0.01]}},
 AxesLabel->{"x",""}];
```

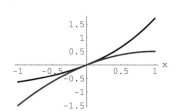

Good transition. Even though you can't make a smooth spline by knotting $f[x]$ and $g[x]$ at $x = 0$, you can make a smooth spline by knotting $F[x] = \int_0^x f[t]\, dt$ and $G[x] = \int_0^x g[t]\, dt$ at $x = 0$.

> Do you think this was an accident? If not, how do you account for it?

■ G.5) Changing the variable to improve the order of contact at 0

G.5.a.i) Put
$$f[x] = e^x \quad \text{and} \quad g[x] = 1 + x + \frac{x^2}{2}.$$
What is the order of contact of $f[x]$ and $g[x]$ at $x = 0$?

G.5.a.ii) Put
$$f[x] = e^x \quad \text{and} \quad g[x] = 1 + x + \frac{x^2}{2}.$$
What is the order of contact of $f[x^2]$ and $g[x^2]$ at $x = 0$?

G.5.a.iii)
> Put
> $$f[x] = e^x \quad \text{and} \quad g[x] = 1 + x + \frac{x^2}{2}.$$
> What is the order of contact of $f[x^3]$ and $g[x^3]$ at $x = 0$?

G.5.b.i)
> If $f[x]$ and $g[x]$ is any pair of functions that have order of contact m at $x = 0$, then what is the order of contact of $f[x^2]$ and $g[x^2]$ at $x = 0$?

G.5.b.ii)
> If $\{f[x], g[x]\}$ is any pair of functions that have order of contact m at $x = 0$, then what is the order of contact of $f[x^3]$ and $g[x^3]$ at $x = 0$?

G.5.b.iii)
> If p is any positive integer and $f[x]$ and $g[x]$ are functions that have order of contact m at $x = 0$, then what is the order of contact of $f[x^p]$ and $g[x^p]$ at $x = 0$?

■ G.6) The natural cubic spline

This problem will appeal mainly to enthusiasts. It is not for everyone.

The natural cubic spline endeavors to put the most artistic curve through a list of points. Here is such a list:

In[14]:=
```
points = {{-8,-12},{-1,-15},{2,20},{5,-4},{8,9},{11,3},{15,9}};
```

There are seven points in this list. One common belief is that the best curve through these points is the unique 6th degree interpolating polynomial through the points. Let's plot it. The equation of this polynomial is:

In[15]:=
```
Clear[x,fitter]
y = Expand[InterpolatingPolynomial[points,x]]
```

Out[15]=
$$\frac{86999543}{6442254} + \frac{612132523\,x}{33611760} - \frac{1637822117\,x^2}{171793440} + \frac{136446911\,x^3}{154614096} + \frac{8114303\,x^4}{77307048} - \frac{577061\,x^5}{32211270} + \frac{974513\,x^6}{1546140960}$$

And the plot:

In[16]:=
```
givenpoints = ListPlot[points,
  PlotStyle->{Red,PointSize[0.025]},
  DisplayFunction->Identity];
polythrough = Plot[y,{x,-8,15},
  DisplayFunction->Identity];
Show[givenpoints,polythrough,
  AxesLabel->{"x","y"},
  DisplayFunction->$DisplayFunction];
```

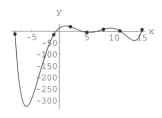

Get real! Look at that huge dip on the left! That polynomial curve has dips and crests that are not suggested by the given points. You'd better give up on using this approach for this list of numbers.

A drastic measure would be to connect the consecutive points with line segments. Some folks call this procedure "linear interpolation." Try it:

In[17]:=
```
sticks = ListPlot[points,PlotJoined->True,
  DisplayFunction->Identity];
Show[givenpoints,sticks,
  AxesLabel->{"x","y"},
  DisplayFunction->$DisplayFunction];
```

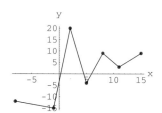

Yuk! This is rough, but it is more satisfactory than the polynomial plot. There must be a way of getting some smoothness into the curve. Enter the natural cubic spline.

Instead of passing a line segment through consecutive points, you pass a different cubic curve through each pair of consecutive points, and make a smooth spline with knots at each of the points. This may seem outlandish at first, because a cubic curve

$$y = a x^3 + b x^2 + c x + d$$

has four coefficients to determine, and normally you fit a cubic through four points. But with the cubic spline, you fit the cubic through just two consecutive points and use the extra freedom you have to guarantee smoothness at the knots on the left and right of the knot. Start out with the points above

$$\{x[1], y[1]\}, \{x[2], y[2]\}, \{x[3], y[3]\}, \{x[4], y[4]\}, \{x[5], y[5]\}, \{x[6], y[6]\}, \{x[7], y[7]\},$$

sorted so that the $x[k]$'s increase as k increases.

In[18]:=
```
Clear[x,y,K];
points = N[Sort[points]];
x[k_] := points[[k,1]];
y[k_] := points[[k,2]];
K = Length[points];
```

$f[k,x]$ is the cubic that you are going to run from $\{x[k], y[k]\}$ to $\{x[k+1], y[k+1]\}$:

In[19]:=
```
Clear[f,a,b,c,d,k,eqn]
f[k_,x_] = a[k] x^3 + b[k] x^2 + c[k] x + d[k]
```
Out[19]=
$$x^3 a[k] + x^2 b[k] + x c[k] + d[k]$$

Here k runs from $k = 1$ to $k = 6$ ($= K - 1$); so you are working with 6 cubics, $f[1,x], f[2,x], \ldots, f[6,x]$ and each cubic has 4 (so far) undetermined coefficients. This means you've got 24 equations to play with.

To hit the points, you want:
$$f[k, x[k]] = y[k] \qquad \text{for } k = 1, 2, 3, 4, 5, 6.$$

In[20]:=
```
eqns1 = Table[f[k,x[k]] == y[k],{k,1,K-1}]
```
Out[20]=
```
{-512. a[1] + 64. b[1] - 8. c[1] + d[1] == -12.,
  -1. a[2] + 1. b[2] - 1. c[2] + d[2] == -15.,
  8. a[3] + 4. b[3] + 2. c[3] + d[3] == 20.,
  125. a[4] + 25. b[4] + 5. c[4] + d[4] == -4.,
  512. a[5] + 64. b[5] + 8. c[5] + d[5] == 9.,
  1331. a[6] + 121. b[6] + 11. c[6] + d[6] == 3.}
```

and
$$f[k, x[k+1]] = y[k+1] \qquad \text{for } k = 1, 2, 3, 4, 5, 6.$$

In[21]:=
```
eqns2 = Table[f[k,x[k+1]] == y[k+1],{k,1,K-1}]
```
Out[21]=
```
{-1. a[1] + 1. b[1] - 1. c[1] + d[1] == -15.,
  8. a[2] + 4. b[2] + 2. c[2] + d[2] == 20.,
  125. a[3] + 25. b[3] + 5. c[3] + d[3] == -4.,
  512. a[4] + 64. b[4] + 8. c[4] + d[4] == 9.,
  1331. a[5] + 121. b[5] + 11. c[5] + d[5] == 3.,
  3375. a[6] + 225. b[6] + 15. c[6] + d[6] == 9.}
```

For smoothness at the knots, you want:
$$f'[k, x[k]] = f'[k-1, x[k]] \qquad \text{for } k = 2, 3, 4, 5, 6$$
and
$$f''[k, x[k]] = f''[k-1, x[k]] \qquad \text{for } k = 2, 3, 4, 5, 6.$$

Here, all derivatives are with respect to the x variable.

In[22]:=
```
eqns3 = Table[(D[f[k,x],x] == D[f[k-1,x],x])/.x->x[k],{k,2,K-1}]
eqns4 = Table[(D[f[k,x],{x,2}] == D[f[k-1,x],{x,2}])/.x->x[k],{k,2,K-1}]
```

Out[22]=
{3. a[2] - 2. b[2] + c[2] == 3. a[1] - 2. b[1] + c[1],
 12. a[3] + 4. b[3] + c[3] == 12. a[2] + 4. b[2] + c[2],
 75. a[4] + 10. b[4] + c[4] == 75. a[3] + 10. b[3] + c[3],
 192. a[5] + 16. b[5] + c[5] == 192. a[4] + 16. b[4] + c[4],
 363. a[6] + 22. b[6] + c[6] == 363. a[5] + 22. b[5] + c[5]}

Out[22]=
{-6. a[2] + 2 b[2] == -6. a[1] + 2 b[1],
 12. a[3] + 2 b[3] == 12. a[2] + 2 b[2],
 30. a[4] + 2 b[4] == 30. a[3] + 2 b[3],
 48. a[5] + 2 b[5] == 48. a[4] + 2 b[4],
 66. a[6] + 2 b[6] == 66. a[5] + 2 b[5]}

You cannot use $k = 1$ in the last two equations because there is no function $f[0, x]$. So far you have specified $6 + 6 + 5 + 5 = 22$ conditions, and you have 24 undetermined coefficients; as a result you need to specify two more conditions. Old-timers at the art of curve fitting have found that good results usually come from specifying that the second derivatives equal 0 at the endpoints, $\{x[1], y[1]\}$ and $\{x[7], y[7]\}$. They call the resulting spline "the natural cubic spline."

In[23]:=
```
eqns5 = {(D[f[1,x],{x,2}]/.x->x[1]) == 0,(D[f[K-1,x],{x,2}]/.x->x[K]) == 0}
```

Out[23]=
{-48. a[1] + 2 b[1] == 0, 90. a[6] + 2 b[6] == 0}

Now you put all of these equations together.

In[24]:=
```
equations = Join[eqns1,eqns2,eqns3,eqns4,eqns5];
```

Solve for the correct coefficients (all 24 of them), substitute, and plot:

In[25]:=
```
coefflist = Flatten[Table[{a[k],b[k],c[k],d[k]},{k,K-1}]];
coeffs = Chop[Solve[equations,coefflist][[1]]]
```

Out[25]=
{a[1] -> 0.136807, b[1] -> 3.28337, c[1] -> 19.1348, d[1] -> 0.988275,
 a[2] -> -1.10341, b[2] -> -0.437284, c[2] -> 15.4142, d[2] -> -0.251943,
 a[3] -> 1.41657, b[3] -> -15.5572, c[3] -> 45.654, d[3] -> -20.4118,
 a[4] -> -1.00732, b[4] -> 20.8011, c[4] -> -136.137, d[4] -> 282.574,
 a[5] -> 0.538617, b[5] -> -16.3013, c[5] -> 160.681, d[5] -> -508.943,
 a[6] -> -0.122758, b[6] -> 5.52411, c[6] -> -79.3976, d[6] -> 371.347}

Now stick all of these coefficients into the formulas for the $f[k, x]$'s and define the natural cubic spline function.

In[26]:=
```
Clear[spline,t,k,g]; g[k_,t_] := f[k,t]/.coeffs;
Do[With[{m = k, n = k + 1},
   spline[t_] := g[m,t]/; x[m] <= t < x[n]],{k, K-1}]
```

Plot it and see what happened:

In[27]:=
```
splineplot = Plot[spline[t],{t,x[1],x[K]},
   PlotRange->All,AxesLabel->{"x","y"}];
```

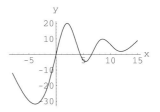

Throw in the points:

In[28]:=
```
Show[splineplot,givenpoints];
```

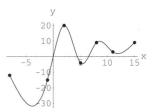

There it is! Golly, that's just about the way most folks would have connected the dots with a pencil. The old-time cubic spliners seem to know what they're talking about. To add adventure to an already thrilling subject, show both the interpolating polynomial and the natural spline on the same plot:

In[29]:=
```
Show[splineplot,polythrough,givenpoints,
   PlotRange->All,AxesLabel->{"x","y"},
   DisplayFunction->$DisplayFunction];
```

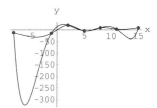

The natural cubic spline shines.

Sit back and reflect, remembering that you are one of an elite group of calculus students who have ever seen a cubic spline.

G.6.a.i) How does the natural cubic spline compare to *Mathematica*'s built in Interpolation function?

Lesson 4.01 Splines

G.6.a.ii) Does this mean that everyone ought to give up on *Mathematica*'s Interpolation function and use the natural cubic spline instead?

G.6.b.i) Take the same points

$$\{-8, -12\}, \{-1, -15\}, \{2, 20\}, \{5, -4\}, \{8, 9\}, \{11, 3\}, \text{ and } \{15, 9\}$$

as above.

Pass a cubic spline through them in the manner of part G.6.a), but this time see what happens when, instead of specifying that the second derivatives be 0 at the endpoints, specify that the first and second derivatives at the left hand endpoint are both equal to 0.

G.6.b.ii) Take the same points

$$\{-8, -12\}, \{-1, -15\}, \{2, 20\}, \{5, -4\}, \{8, 9\}, \{11, 3\}, \text{ and } \{15, 9\}$$

as above.

Pass a cubic spline through them in the manner of part a), but this time see what happens when, instead of specifying that the second derivatives be 0 at the endpoints, specify the values to be other numbers.

Try positive numbers, negative numbers, or mixtures.

G.6.c) Pass the unique eighth degree polynomial through the following data points:

```
In[30]:=
  points = {{0.2,0.9},{0.4,1.8},{0.6,4.6},{0.8,10.8},{1.0,5.},
  {1.2,2.6},{1.4,2.1},{1.6,2.},{1.8,2.1}};
```

And pass the natural cubic spline through the same data points. Plot the polynomial and the spline on separate plots, and discuss the virtues of each.

LESSON 4.02

Expansions in Powers of x

Basics

■ **B.1) The expansion of a function $f[x]$ in powers of x**

Given a function $f[x]$, the expansion of $f[x]$ in powers of x is

$$a[0] + a[1]\,x + a[2]\,x^2 + a[3]\,x^3 + \cdots + a[k]\,x^k + a[k+1]\,x^{k+1} + \cdots$$

where the numbers $a[0], a[1], a[2], \ldots, a[k], a[k+1], \ldots$ are chosen so that for every positive integer m, the function $f[x]$ and the polynomial

$$a[0] + a[1]\,x + a[2]\,x^2 + a[3]\,x^3 + \cdots + a[m]\,x^m$$

have order of contact m at $x = 0$. That's quite a pill to swallow, but the idea is not all that hard once you've gotten some experience.

B.1.a) Take $f[x] = 1/(2+x)$. To get the expansion of $f[x]$ in powers of x, you could try to work out the mth degree polynomial that has order of contact m with $f[x]$ at $x = 0$ as you did in the previous lesson on splines. But *Mathematica* has a built-in instruction that will do this work automatically for you. Here is the second degree polynomial that has order of contact 2 with $f[x]$ at $x = 0$:

In[1]:=
```
Clear[f,x]; f[x_] = 1/(2 + x);
Normal[Series[f[x],{x,0,2}]]
```

Out[1]=

$$\frac{1}{2} - \frac{x}{4} + \frac{x^2}{8}$$

Lesson 4.02 Expansions in Powers of x

Here is the third degree polynomial that has order of contact 3 with $f[x]$ at $x = 0$:

In[2]:=
```
Normal[Series[f[x],{x,0,3}]]
```

Out[2]=
$$\frac{1}{2} - \frac{x}{4} + \frac{x^2}{8} - \frac{x^3}{16}$$

Here is the fourth degree polynomial that has order of contact 4 with $f[x]$ at $x = 0$:

In[3]:=
```
Normal[Series[f[x],{x,0,4}]]
```

Out[3]=
$$\frac{1}{2} - \frac{x}{4} + \frac{x^2}{8} - \frac{x^3}{16} + \frac{x^4}{32}$$

Here is the tenth degree polynomial that has order of contact 10 with $f[x]$ at $x = 0$:

In[4]:=
```
Normal[Series[f[x],{x,0,10}]]
```

Out[4]=
$$\frac{1}{2} - \frac{x}{4} + \frac{x^2}{8} - \frac{x^3}{16} + \frac{x^4}{32} - \frac{x^5}{64} + \frac{x^6}{128} - \frac{x^7}{256} + \frac{x^8}{512} - \frac{x^9}{1024} + \frac{x^{10}}{2048}$$

> Look at the pattern emerging above, and give the expansion of $f[x] = 1/(2 + x)$ in powers of x.

Answer: Look again:

In[5]:=
```
Table[Normal[Series[f[x],{x,0,k}]],{k,3,6}]//TableForm
```

Out[5]=
$$\frac{1}{2} - \frac{x}{4} + \frac{x^2}{8} - \frac{x^3}{16}$$

$$\frac{1}{2} - \frac{x}{4} + \frac{x^2}{8} - \frac{x^3}{16} + \frac{x^4}{32}$$

$$\frac{1}{2} - \frac{x}{4} + \frac{x^2}{8} - \frac{x^3}{16} + \frac{x^4}{32} - \frac{x^5}{64}$$

$$\frac{1}{2} - \frac{x}{4} + \frac{x^2}{8} - \frac{x^3}{16} + \frac{x^4}{32} - \frac{x^5}{64} + \frac{x^6}{128}$$

Those denominators are consecutive powers of 2, and the signs alternate. The expansion of $f[x] = 1/(2+x)$ in powers of x is

$$\frac{1}{2} - \frac{x}{2^2} + \frac{x^2}{2^3} - \frac{x^3}{2^4} + \frac{x^4}{2^5} \cdots + \frac{(-1)^k x^k}{2^{k+1}} + \cdots$$

This is

$$a[0] + a[1]\,x + a[2]\,x^2 + a[3]\,x^3 + \cdots + a[k]\,x^k + a[k+1]\,x^{k+1} + \cdots$$

with $a[k] = (-1)^k / 2^{k+1}$. Try it out:

In[6]:=
```
Clear[a,k]
a[k_] = ((-1)^k)/2^(k + 1);
Table[Sum[a[k]x^k,{k,0,m}],{m,3,6}]//TableForm
```

Out[6]=
```
                 2    3
1    x    x    x
- -  - +  - -  --
2    4    8    16

                 2    3    4
1    x    x    x    x
- -  - +  - -  - +  --
2    4    8    16   32

                 2    3    4    5
1    x    x    x    x    x
- -  - +  - -  - +  - -  --
2    4    8    16   32   64

                 2    3    4    5    6
1    x    x    x    x    x    x
- -  - +  - -  - +  - -  - +  --
2    4    8    16   32   64   128
```

It checks.

■ B.2) The expansions every literate calculus person knows

B.2.a) Look at these:

In[7]:=
```
Clear[f,x]; f[x_] = 1/(1 - x); b = 0;
Normal[Series[f[x],{x,b,2}]]
```

Out[7]=
```
         2
1 + x + x
```

In[8]:=
```
Normal[Series[f[x],{x,b,3}]]
```

Out[8]=
```
         2    3
1 + x + x  + x
```

In[9]:=
```
Normal[Series[f[x],{x,b,4}]]
```

Lesson 4.02 Expansions in Powers of x

Out[9]=
$$1 + x + x^2 + x^3 + x^4$$

In[10]:=
```
Normal[Series[f[x],{x,b,12}]]
```

Out[10]=
$$1 + x + x^2 + x^3 + x^4 + x^5 + x^6 + x^7 + x^8 + x^9 + x^{10} + x^{11} + x^{12}$$

> Look at the pattern emerging above and give the expansion of
> $$f[x] = \frac{1}{1-x}$$
> in powers of x.

Answer: Look again:

In[11]:=
```
Normal[Series[f[x],{x,b,50}]]
```

Out[11]=
$$1 + x + x^2 + x^3 + x^4 + x^5 + x^6 + x^7 + x^8 + x^9 + x^{10} + x^{11} + x^{12} + x^{13} + x^{14} + x^{15} +$$
$$x^{16} + x^{17} + x^{18} + x^{19} + x^{20} + x^{21} + x^{22} + x^{23} + x^{24} + x^{25} + x^{26} + x^{27} + x^{28} +$$
$$x^{29} + x^{30} + x^{31} + x^{32} + x^{33} + x^{34} + x^{35} + x^{36} + x^{37} + x^{38} + x^{39} + x^{40} + x^{41} +$$
$$x^{42} + x^{43} + x^{44} + x^{45} + x^{46} + x^{47} + x^{48} + x^{49} + x^{50}$$

Ain't no doubt about it. The expansion of $f[x] = 1/(1-x)$ in powers of x is
$$1 + x + x^2 + x^3 + x^4 + \cdots + x^k + \cdots$$

This is
$$a[0] + a[1]\,x + a[2]\,x^2 + a[3]\,x^3 + a[4]\,x^4 + \cdots + a[k]\,x^k + a[k+1]\,x^{k+1} + \cdots$$
with $a[k] = 1$ for all the k's.

Know this cold!

B.2.b) Look at these:

In[12]:=
```
Clear[f,x]; f[x_] = E^x; b = 0;
Normal[Series[f[x],{x,b,2}]]
```

Out[12]=
$$1 + x + \frac{x^2}{2}$$

In[13]:=
```
Normal[Series[f[x],{x,b,3}]]
```

Out[13]=
$$1 + x + \frac{x^2}{2} + \frac{x^3}{6}$$

In[14]:=
`Normal[Series[f[x],{x,b,4}]]`

Out[14]=
$$1 + x + \frac{x^2}{2} + \frac{x^3}{6} + \frac{x^4}{24}$$

In[15]:=
`Normal[Series[f[x],{x,b,6}]]`

Out[15]=
$$1 + x + \frac{x^2}{2} + \frac{x^3}{6} + \frac{x^4}{24} + \frac{x^5}{120} + \frac{x^6}{720}$$

Look at the pattern emerging above, and give the expansion of $f[x] = e^x$ in powers of x.

Answer: Look again:

In[16]:=
`Normal[Series[f[x],{x,b,8}]]`

Out[16]=
$$1 + x + \frac{x^2}{2} + \frac{x^3}{6} + \frac{x^4}{24} + \frac{x^5}{120} + \frac{x^6}{720} + \frac{x^7}{5040} + \frac{x^8}{40320}$$

Those denominators are factorials.

In[17]:=
`Clear[k]; Table[k!,{k,0,8}]`

Out[17]=
{1, 1, 2, 6, 24, 120, 720, 5040, 40320}

Ain't no doubt about it. The expansion of $f[x] = e^x$ in powers of x is

$$1 + x + \frac{x^2}{2!} + \frac{x^3}{3!} + \frac{x^4}{4!} + \cdots + \frac{x^k}{k!} + \cdots$$

This is

$$a[0] + a[1]\,x + a[2]\,x^2 + a[3]\,x^3 + a[4]\,x^4 + \cdots + a[k]\,x^k + a[k+1]\,x^{k+1} + \cdots$$

with $a[k] = 1/k!$ for all the k's.

Know this cold!

You might not be used to saying that $0! = 1$, but *Mathematica* says this:

Lesson 4.02 Expansions in Powers of x

In[18]:=
```
0!
```
Out[18]=
1

And so do all good folks.

B.2.c) Look at these:

In[19]:=
```
Clear[f,x]; f[x_] = Sin[x]; b = 0;
Normal[Series[f[x],{x,b,3}]]
```
Out[19]=
$$x - \frac{x^3}{6}$$

In[20]:=
```
Normal[Series[f[x],{x,b,5}]]
```
Out[20]=
$$x - \frac{x^3}{6} + \frac{x^5}{120}$$

In[21]:=
```
Normal[Series[f[x],{x,b,7}]]
```
Out[21]=
$$x - \frac{x^3}{6} + \frac{x^5}{120} - \frac{x^7}{5040}$$

> Look at the pattern emerging above, and give the expansion of $f[x] = \sin[x]$ in powers of x.

Answer: Look again:

In[22]:=
```
Normal[Series[f[x],{x,b,17}]]
```
Out[22]=
$$x - \frac{x^3}{6} + \frac{x^5}{120} - \frac{x^7}{5040} + \frac{x^9}{362880} - \frac{x^{11}}{39916800} + \frac{x^{13}}{6227020800} - \frac{x^{15}}{1307674368000} + \frac{x^{17}}{355687428096000}$$

Those denominators are factorials of consecutive odd integers. And the signs alternate.

In[23]:=
```
Clear[k]; Table[(2 k + 1)!,{k,0,8}]
```

Out[23]=
{1, 6, 120, 5040, 362880, 39916800, 6227020800, 1307674368000, 355687428096000}

Not much doubt about it. The expansion of $f[x] = \sin[x]$ in powers of x is
$$x - \frac{x^3}{3!} + \frac{x^5}{5!} - \frac{x^7}{7!} + \cdots + \frac{(-1)^k x^{2k+1}}{(2k+1)!} + \cdots$$
Odd powers because $\sin[x]$ is an odd function.

This is
$$a[0] + a[1]\,x + a[2]\,x^2 + a[3]\,x^3 + a[4]\,x^4 + \cdots + a[k]\,x^k + a[k+1]\,x^{k+1} + \cdots$$
with $a[2k] = 0$ and $a[2k+1] = (-1)^k / (2k+1)!$.

Know this!

B.2.d) Look at these:

In[24]:=
```
Clear[f,x]; f[x_] = Cos[x]; b = 0;
Normal[Series[f[x],{x,b,2}]]
```
Out[24]=
$$1 - \frac{x^2}{2}$$

In[25]:=
```
Normal[Series[f[x],{x,b,4}]]
```
Out[25]=
$$1 - \frac{x^2}{2} + \frac{x^4}{24}$$

In[26]:=
```
Normal[Series[f[x],{x,b,6}]]
```
Out[26]=
$$1 - \frac{x^2}{2} + \frac{x^4}{24} - \frac{x^6}{720}$$

Look at the pattern emerging above, and give the expansion of $f[x] = \cos[x]$ in powers of x.

Answer: Look again:

In[27]:=
```
Normal[Series[f[x],{x,b,16}]]
```
Out[27]=
$$1 - \frac{x^2}{2} + \frac{x^4}{24} - \frac{x^6}{720} + \frac{x^8}{40320} - \frac{x^{10}}{3628800} + \frac{x^{12}}{479001600} - \frac{x^{14}}{87178291200} + \frac{x^{16}}{20922789888000}$$

Those denominators are factorials of consecutive even integers. And the signs alternate.

In[28]:=
```
Clear[k]; Table[(2 k)!,{k,0,8}]
```
Out[28]=
{1, 2, 24, 720, 40320, 3628800, 479001600, 87178291200, 20922789888000}

Not much doubt about it. The expansion of $f[x] = \cos[x]$ in powers of x is

$$1 - \frac{x^2}{2!} + \frac{x^4}{4!} - \frac{x^6}{6!} + \cdots + \frac{(-1)^k x^{2k}}{(2k)!} + \cdots$$

Even powers because $\cos[x]$ is an even function.

This is

$$a[0] + a[1] x + a[2] x^2 + a[3] x^3 + a[4] x^4 + \cdots + a[k] x^k + a[k+1] x^{k+1} + \cdots$$

with $a[2k+1] = 0$ and $a[2k] = (-1)^k / (2k)!$.

Know this!

■ B.3) Expansions for approximation

Go with $f[x] = \sin[x]$ whose expansion in powers of x is

$$x - \frac{x^3}{3!} + \frac{x^5}{5!} + \frac{x^7}{7!} - \frac{x^9}{9!} + \cdots + \frac{(-1)^k x^{2k+1}}{(2k+1)!} + \cdots.$$

Here's what you get when you plot $f[x]$ and the polynomial made from using the terms through the x^3 term of its expansion in powers of x:

In[29]:=
```
Clear[f,approx3,x,m]; f[x_] = Sin[x];
approx3[x_] = Normal[Series[f[x],{x,0,3}]]
```
Out[29]=
$$x - \frac{x^3}{6}$$

In[30]:=
```
Plot[{f[x],approx3[x]},{x,-4,4},
 PlotStyle->{{Blue,Thickness[0.02]},
 {Red,Thickness[0.01]}},
 AxesLabel->{"x",""},PlotRange->{-2,2},
 PlotLabel->"Sin[x] and" approx3[x]];
```

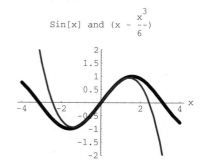

Sharing some ink.

Here's what you get when you plot $f[x]$ and the polynomial made from using the terms through the x^5 term of its expansion in powers of x:

In[31]:=
```
Clear[f,approx5,x,m]; f[x_] = Sin[x];
approx5[x_] = Normal[Series[f[x],{x,0,5}]]
```

Out[31]=
$$x - \frac{x^3}{6} + \frac{x^5}{120}$$

In[32]:=
```
Plot[{f[x],approx5[x]},{x,-4,4},
PlotStyle->{{Blue,Thickness[0.02]},
{Red,Thickness[0.01]}},
AxesLabel->{"x",""},PlotRange->{-2,2},
PlotLabel->"Sin[x] and" approx5[x]];
```

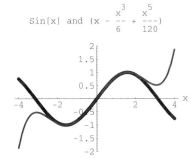

Sharing more ink.

Here's what you get when you plot $f[x]$ and the polynomial made from using the terms through the x^7 term of its expansion in powers of x:

In[33]:=
```
Clear[f,approx7,x,m]; f[x_] = Sin[x];
approx7[x_] = Normal[Series[f[x],{x,0,7}]]
```

Out[33]=
$$x - \frac{x^3}{6} + \frac{x^5}{120} - \frac{x^7}{5040}$$

In[34]:=
```
Plot[{f[x],approx7[x]},{x,-4,4},
PlotStyle->{{Blue,Thickness[0.02]},
{Red,Thickness[0.01]}},
AxesLabel->{"x",""},PlotRange->{-2,2},
PlotLabel->"Sin[x] and" approx7[x]];
```

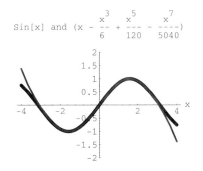

Sharing even more ink.

Lesson 4.02 Expansions in Powers of x

Here's what you get when you plot $f[x]$ and the polynomial made from using the terms through the x^9 term of its expansion in powers of x:

In[35]:=
```
Clear[f,approx9,x,m]; f[x_] = Sin[x];
approx9[x_] = Normal[Series[f[x],{x,0,9}]]
```

Out[35]=
$$x - \frac{x^3}{6} + \frac{x^5}{120} - \frac{x^7}{5040} + \frac{x^9}{362880}$$

In[36]:=
```
Plot[{f[x],approx9[x]},{x,-4,4},
PlotStyle->{{Blue,Thickness[0.02]},
{Red,Thickness[0.01]}},
AxesLabel->{"x",""},PlotRange->{-2,2},
PlotLabel->"Sin[x] and" approx9[x]];
```

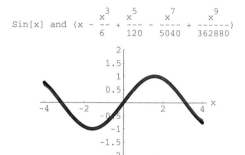

Sharing ink all the way.

B.3.a) Why did this happen?

Answer: The facts of the matter:

→ As you use more and more terms from the expansion in powers of x, you increase the order of contact between $f[x]$ and the corresponding polynomial at $x = 0$.

→ As you increase the order of contact at $x = 0$, you increase the quality of the transition from one curve to the other at $x = 0$.

Evidently, when the transition is really good, the curves have no choice but to share a lot of ink.

B.3.b) What good is this?

Answer: It's good for approximations. In real life, no one knows how to calculate the exact value of $\sin[x]$ for all x's between $-\pi/2$ and $\pi/2$. Instead, serious number crunchers rely on approximations. Take a gander at this:

In[37]:=
```
Clear[f,approx13,x,m]; f[x_] = Sin[x];
approx13[x_] = Normal[Series[f[x],{x,0,13}]]
```

Out[37]=
$$x - \frac{x^3}{6} + \frac{x^5}{120} - \frac{x^7}{5040} + \frac{x^9}{362880} - \frac{x^{11}}{39916800} + \frac{x^{13}}{6227020800}$$

In[38]:=
```
Plot[{f[x],approx13[x]},{x,-Pi/2,Pi/2},
 PlotStyle->{{Blue,Thickness[0.02]},
  {Red,Thickness[0.01]}},
 AxesLabel->{"x",""}];
```

Running together all the way. See how close they really are by plotting their difference for $-\pi/2 \le x \le \pi/2$:

In[39]:=
```
Plot[f[x] - approx13[x],{x,-Pi/2,Pi/2},
 PlotStyle->Red,PlotRange->All,
 AxesLabel->{"x",""}];
```

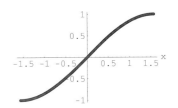

The upshot: For $-\pi/2 \le x \le \pi/2$, $f[x] = \sin[x]$ is equal to approx13[x] to within an error of no more than 10^{-9}.

If you have a calculator that can only add, subtract, multiply, and divide, then you can calculate $\sin[x]$ for $-\pi/2 \le x \le \pi/2$ damn accurately by calculating:

In[40]:=
```
approx13[x]
```

Out[40]=
$$x - \frac{x^3}{6} + \frac{x^5}{120} - \frac{x^7}{5040} + \frac{x^9}{362880} - \frac{x^{11}}{39916800} + \frac{x^{13}}{6227020800}$$

Pretty nifty.

B.3.c) | What's the best way to think of expansions?

Answer: Expansions are clay for approximations. You might never use the whole expansion in powers of x of a function $f[x]$, but usually you can use enough of it to approximate $f[x]$ very well on an interval centered at 0. Try it:

In[41]:=
```
Clear[f,approx3,x,m]; f[x_] = E^x Sin[3 x];
approx3[x_] = Normal[Series[f[x],{x,0,3}]]
```

Out[41]=
$$3x + 3x^2 - 3x^3$$

In[42]:=
```
Plot[{f[x],approx3[x]},{x,-1,1},
 PlotStyle->{{Blue,Thickness[0.02]},
 {Red,Thickness[0.01]}},
 AxesLabel->{"x",""}];
```

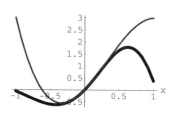

If you want a better approximation, then you use more of the expansion:

In[43]:=
```
Clear[approx6]
approx6[x_] = Normal[Series[f[x],{x,0,6}]]
```

Out[43]=
$$3x + 3x^2 - 3x^3 - 4x^4 - \frac{x^5}{10} + \frac{13x^6}{10}$$

In[44]:=
```
Plot[{f[x],approx6[x]},{x,-1, 1},
 PlotStyle->{{Blue,Thickness[0.02]},
 {Red,Thickness[0.01]}},
 AxesLabel->{"x",""}];
```

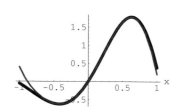

Math happens again.

Tutorials

■ T.1) Expansions by substitution

Use the basic expansions

→ $1/(1-x)$ in powers of x:
$$1 + x + x^2 + x^3 + x^4 + \cdots + x^k + \cdots$$

→ e^x in powers of x:
$$1 + x + \frac{x^2}{2!} + \frac{x^3}{3!} + \cdots + \frac{x^k}{k!} + \cdots$$

→ $\sin[x]$ in powers of x:
$$x - \frac{x^3}{3!} + \frac{x^5}{5!} + \cdots + \frac{(-1)^k x^{2k+1}}{(2k+1)!} + \cdots$$

$\rightarrow \cos[x]$ in powers of x:

$$1 - \frac{x^2}{2!} + \frac{x^4}{4!} + \cdots + \frac{(-1)^k x^{2k}}{(2k)!} + \cdots$$

to arrive at the expansions in powers of x of the following functions $f[x]$. Use *Mathematica* to check yourself.

T.1.a) $\quad f[x] = e^{x^2}$

Answer: The expansion of e^x in powers of x is

$$1 + x + \frac{x^2}{2!} + \frac{x^3}{3!} + \frac{x^4}{4!} + \frac{x^5}{5!} + \cdots + \frac{x^k}{k!} + \cdots$$

To get the expansion of e^{x^2} in powers of x, just replace x by x^2 throughout; this gives the expansion

$$1 + x^2 + \frac{x^4}{2!} + \frac{x^6}{3!} + \frac{x^8}{4!} + \frac{x^{10}}{5!} + \cdots + \frac{x^{2k}}{k!} + \cdots$$

of e^{x^2} in powers of x. Run this through *Mathematica*:

In[1]:=
```
Clear[x]; Normal[Series[E^x,{x,0,8}]]
```
Out[1]=
$$1 + x + \frac{x^2}{2} + \frac{x^3}{6} + \frac{x^4}{24} + \frac{x^5}{120} + \frac{x^6}{720} + \frac{x^7}{5040} + \frac{x^8}{40320}$$

In[2]:=
```
Normal[Series[E^x,{x,0,8}]]/.x->x^2
```
Out[2]=
$$1 + x^2 + \frac{x^4}{2} + \frac{x^6}{6} + \frac{x^8}{24} + \frac{x^{10}}{120} + \frac{x^{12}}{720} + \frac{x^{14}}{5040} + \frac{x^{16}}{40320}$$

In[3]:=
```
Normal[Series[E^(x^2),{x,0,16}]]
```
Out[3]=
$$1 + x^2 + \frac{x^4}{2} + \frac{x^6}{6} + \frac{x^8}{24} + \frac{x^{10}}{120} + \frac{x^{12}}{720} + \frac{x^{14}}{5040} + \frac{x^{16}}{40320}$$

Right on the money.

T.1.b) $\quad f[x] = \sin[x^3]$

Lesson 4.02 Expansions in Powers of x

Answer: The expansion of $\sin[x]$ in powers of x is

$$x - \frac{x^3}{3!} + \frac{x^5}{5!} - \frac{x^7}{7!} + \cdots + \frac{(-1)^k x^{2k+1}}{(2k+1)!} + \cdots$$

To get the expansion of $\sin[x^3]$ in powers of x, just replace x by x^3 throughout. This gives the expansion

$$x^3 - \frac{x^9}{3!} + \frac{x^{15}}{5!} - \frac{x^{21}}{7!} + \cdots + \frac{(-1)^k x^{6k+3}}{(2k+1)!} + \cdots$$

of $\sin[x^3]$ in powers of x. Check:

In[4]:=
```
Clear[x]; Normal[Series[Sin[x],{x,0,13}]]
```
Out[4]=
$$x - \frac{x^3}{6} + \frac{x^5}{120} - \frac{x^7}{5040} + \frac{x^9}{362880} - \frac{x^{11}}{39916800} + \frac{x^{13}}{6227020800}$$

In[5]:=
```
Normal[Series[Sin[x],{x,0,13}]]/.x->x^3
```
Out[5]=
$$x^3 - \frac{x^9}{6} + \frac{x^{15}}{120} - \frac{x^{21}}{5040} + \frac{x^{27}}{362880} - \frac{x^{33}}{39916800} + \frac{x^{39}}{6227020800}$$

In[6]:=
```
Normal[Series[Sin[x^3],{x,0,39}]]
```
Out[6]=
$$x^3 - \frac{x^9}{6} + \frac{x^{15}}{120} - \frac{x^{21}}{5040} + \frac{x^{27}}{362880} - \frac{x^{33}}{39916800} + \frac{x^{39}}{6227020800}$$

It checks.

T.1.c)
$$f[x] = \frac{1}{1-x^2}$$

Answer: The expansion of $1/(1-x)$ in powers of x is

$$1 + x + x^2 + x^3 + \cdots + x^k + \cdots$$

To get the expansion of $1/(1-x^2)$ in powers of x just replace x by x^2 throughout. This gives the expansion

$$1 + x^2 + x^4 + x^6 + \cdots + x^{2k} + \cdots$$

This should be the expansion of $1/(1-x^2)$ in powers of x. Run it through *Mathematica*:

In[7]:=
```
Clear[x]; Normal[Series[1/(1 - x),{x,0,10}]]
```
Out[7]=
$$1 + x + x^2 + x^3 + x^4 + x^5 + x^6 + x^7 + x^8 + x^9 + x^{10}$$

In[8]:=
```
Normal[Series[1/(1 - x),{x,0,10}]]/.x->x^2
```
Out[8]=
$$1 + x^2 + x^4 + x^6 + x^8 + x^{10} + x^{12} + x^{14} + x^{16} + x^{18} + x^{20}$$

In[9]:=
```
Normal[Series[1/(1 - x^2),{x,0,20}]]
```
Out[9]=
$$1 + x^2 + x^4 + x^6 + x^8 + x^{10} + x^{12} + x^{14} + x^{16} + x^{18} + x^{20}$$

It checks out.

T.1.d) $$f[x] = \frac{x}{1+x^4}$$

Answer: You go from $1/(1-x)$ to $x/(1+x^4)$ by changing x to $-x^4$ and then multiplying by x. The expansion of $1/(1-x)$ in powers of x is

$$1 + x + x^2 + x^3 + \cdots + x^k + \cdots$$

To get the expansion of $1/(1+x^4)$ in powers of x, just replace x by $-x^4$ throughout; this gives the expansion

$$1 - x^4 + x^8 - x^{12} + \cdots + (-1)^k x^{4k} + \cdots$$

Now multiply by x to get

$$x - x^5 + x^9 - x^{13} + \cdots + (-1)^k x^{4k+1} + \cdots$$

This should be the expansion of $x/(1+x^4)$ in powers of x. Run it through *Mathematica*:

In[10]:=
```
step1 = Normal[Series[1/(1 - x),{x,0,10}]]/.x->-x^4
```
Out[10]=
$$1 - x^4 + x^8 - x^{12} + x^{16} - x^{20} + x^{24} - x^{28} + x^{32} - x^{36} + x^{40}$$

In[11]:=
```
step2 = Expand[x step1]
```
Out[11]=
$$x - x^5 + x^9 - x^{13} + x^{17} - x^{21} + x^{25} - x^{29} + x^{33} - x^{37} + x^{41}$$

In[12]:=
```
Normal[Series[x/(1 + x^4),{x,0,41}]]
```

Out[12]=
$$x - x^5 + x^9 - x^{13} + x^{17} - x^{21} + x^{25} - x^{29} + x^{33} - x^{37} + x^{41}$$

Perfect.

■ T.2) Expansions by differentiation

T.2.a.i) Look at this:

In[13]:=
```
Clear[f,x]; f[x_] = 1/(1 + x);
Normal[Series[f[x],{x,0,8}]]
```

Out[13]=
$$1 - x + x^2 - x^3 + x^4 - x^5 + x^6 - x^7 + x^8$$

And look at this:

In[14]:=
```
Normal[Series[f'[x],{x,0,7}]]
```

Out[14]=
$$-1 + 2x - 3x^2 + 4x^3 - 5x^4 + 6x^5 - 7x^6 + 8x^7$$

What's the message?

Answer: Look again:

In[15]:=
```
Clear[f,x]; f[x_] = 1/(1 + x);
first = Normal[Series[f[x],{x,0,11}]]
```

Out[15]=
$$1 - x + x^2 - x^3 + x^4 - x^5 + x^6 - x^7 + x^8 - x^9 + x^{10} - x^{11}$$

In[16]:=
```
second = Normal[Series[f'[x],{x,0,10}]]
```

Out[16]=
$$-1 + 2x - 3x^2 + 4x^3 - 5x^4 + 6x^5 - 7x^6 + 8x^7 - 9x^8 + 10x^9 - 11x^{10}$$

In[17]:=
```
second == D[first,x]
```

Out[17]=
True

The message is that if you have the expansion of $f[x]$ in powers of x, then you can get the expansion of $f'[x]$ in powers of x simply by differentiating the expansion of $f[x]$. When you think about it, you will agree that this is automatic.

Reason:

→ If
$$a[0] + a[1]\,x + a[2]\,x^2 + a[3]\,x^3 + \cdots + a[m]\,x^m$$
is the sum of the first m terms of the expansion of $f[x]$ in powers of x, then this polynomial has order of contact m with $f[x]$ at $x = 0$.

→ When you differentiate both, you lose one order of contact. So
$$a[1] + 2\,a[2]\,x + 3\,a[3]\,x^2 + \cdots + m\,a[m]\,x^{m-1}$$
is a polynomial of degree $(m-1)$ which has order of contact $(m-1)$ with $f'[x]$ at $x = 0$.

Because this works for any m, this is enough to guarantee that you can get the expansion of $f'[x]$ in powers of x simply by differentiating the expansion of $f[x]$ in powers of x.

T.2.a.ii) The expansion of $1/(1-x)$ in powers of x is
$$1 + x + x^2 + x^3 + x^4 + x^5 + \cdots + x^k + \cdots$$

The derivative of
$$\frac{1}{1-x} \quad \text{is} \quad \frac{1}{(1-x)^2}:$$

In[18]:=
```
Clear[x]; D[1/(1 - x),x]
```
Out[18]=
```
      -2
(1 - x)
```

Use these facts to help write down the expansion of $1/(1-x)^2$ in powers of x.

Answer: Just take the expansion of $1/(1-x)$ in powers of x
$$1 + x + x^2 + x^3 + x^4 + x^5 + x^6 \cdots + x^k + \cdots$$
and differentiate it to get
$$1 + 2\,x + 3\,x^2 + 4\,x^3 + 5\,x^4 + \cdots + k\,x^{k-1} + \cdots$$

Check:

In[19]:=
```
Normal[Series[1/(1 - x)^2,{x,0,10}]]
```
Out[19]=
```
            2      3      4      5      6      7      8      9       10
1 + 2 x + 3 x  + 4 x  + 5 x  + 6 x  + 7 x  + 8 x  + 9 x  + 10 x  + 11 x
```

Looks fine.

■ T.3) Expansions by integration

T.3.a.i) Look at this:

In[20]:=
```
Clear[f,x]; f[x_] = Cos[x];
Normal[Series[f[x],{x,0,8}]]
```

Out[20]=
$$1 - \frac{x^2}{2} + \frac{x^4}{24} - \frac{x^6}{720} + \frac{x^8}{40320}$$

And look at this:

In[21]:=
```
Clear[F,t]; F[x_] = Integrate[f[t],{t,0,x}]
```

Out[21]=
Sin[x]

In[22]:=
```
Normal[Series[F[x],{x,0,9}]]
```

Out[22]=
$$x - \frac{x^3}{6} + \frac{x^5}{120} - \frac{x^7}{5040} + \frac{x^9}{362880}$$

> What's the message?

Answer: Look again, remembering that $F[x] = \int_0^x f[t]\,dt$:

In[23]:=
```
first = Normal[Series[f[x],{x,0,11}]]
```

Out[23]=
$$1 - \frac{x^2}{2} + \frac{x^4}{24} - \frac{x^6}{720} + \frac{x^8}{40320} - \frac{x^{10}}{3628800}$$

In[24]:=
```
second = Normal[Series[F[x],{x,0,12}]]
```

Out[24]=
$$x - \frac{x^3}{6} + \frac{x^5}{120} - \frac{x^7}{5040} + \frac{x^9}{362880} - \frac{x^{11}}{39916800}$$

In[25]:=
```
first == D[second,x]
```

Out[25]=
True

The message is that if you have the expansion of $f[x]$ in powers of x, then you can get the expansion of $F[x] = \int_0^x f[t]\,dt$ in powers of x simply by integrating the expansion of $f[x]$ from 0 to x.

When you think about it, this is automatic.

Reason:

→ If
$$a[0] + a[1]\,x + a[2]\,x^2 + a[3]\,x^3 + \cdots + a[m]\,x^m$$
is the sum of the first m terms of the expansion of $f[x]$ in powers of x, then this polynomial has order of contact m with $f[x]$ at $x = 0$.

→ When you integrate both from b to x, you gain one order of contact at $x = 0$. So
$$a[0]\,x + \frac{a[1]\,x^2}{2} + \frac{a[2]\,x^3}{3} + \cdots + \frac{a[m]\,x^{m+1}}{m+1}$$
is a polynomial of degree $m+1$ which has order of contact $(m+1)$ with $F[x] = \int_0^x f[t]\,dt$ at $x = 0$.

Because this works for any m, this is enough to guarantee that you can get the expansion of $F[x]$ in powers of x simply by integrating the expansion of $f[x]$ in powers of x from 0 to x.

T.3.a.ii) | Give the expansion of $\log[1 + x^2]$ in powers of x.

Answer: You are probably clueless on this, but the purpose of a Tutorial problem is to clue you in. Here you go. Take the derivative of $\log[1 + x^2]$ with the hope that the derivative has an easy expansion:

In[26]:=
  Clear[x]; D[Log[1 + x^2],x]

Out[26]=
$$\frac{2\,x}{1 + x^2}$$

Now integrate back:

In[27]:=
  Clear[x,t]; Integrate[2 t/(1 + t^2),{t,0,x}]

Out[27]=
  Log[1 + x^2]

This tells you that you can integrate the expansion of $2t/(1+t^2)$ in powers of t from 0 to x to get the expansion of $\log[1+x^2]$ in powers of x. Go for it:

Lesson 4.02 Expansions in Powers of x

The expansion of $1/(1-t)$ in powers of t is

$$1 + t + t^2 + t^3 + \cdots + t^k + \cdots$$

To get the expansion of $1/(1+t^2)$, just replace t by $-t^2$ throughout; this gives the expansion

$$1 - t^2 + t^4 - t^6 + \cdots + (-1)^k t^{2k} + \cdots$$

To get the expansion of $2t/(1+t^2)$, multiply by $2t$ to get

$$2t - 2t^3 + 2t^5 - 2t^7 + \cdots + (-1)^k 2t^{2k+1} + \cdots$$

This is the expansion of $2t/(1+t^2)$ in powers of t. Because

$$\log[1+x^2] = \int_0^x \frac{2t}{1+t^2}\,dt,$$

this tells you that the expansion of $\log[1+x^2]$ in powers of x is

$$\frac{2x^2}{2} - \frac{2x^4}{4} + \frac{2x^6}{6} + \frac{2x^8}{8} + \cdots + \frac{(-1)^k 2x^{2k+2}}{(2k+2)} + \cdots$$

$$= x^2 - \frac{x^4}{2} + \frac{x^6}{3} + \frac{x^8}{4} + \cdots + \frac{(-1)^k x^{2k+2}}{(k+1)} + \cdots$$

Check it out:

In[28]:=
```
Clear[x]; Normal[Series[Log[1 + x^2],{x,0,12}]]
```

Out[28]=
$$x^2 - \frac{x^4}{2} + \frac{x^6}{3} - \frac{x^8}{4} + \frac{x^{10}}{5} - \frac{x^{12}}{6}$$

Looking just fine, thank you.

Give It a Try

Experience with the starred $(\star)$ problems will be especially beneficial for understanding later lessons.

■ G.1) A festival of expansions and approximations*

In the following plotting experiments, you are not asked to find the general form of any expansion. Use *Mathematica* to deliver any part of the expansion you happen to need.

G.1.a) Go with $f[x] = \cos[x]$. Here is a plot of $f[x]$ and the polynomial made from using the the terms through the x^2 term of its expansion in powers of x:

In[1]:=
```
Clear[f,approx2,x]; f[x_] = Cos[x];
approx2[x_] = Normal[Series[f[x],{x,0,2}]]
```

Out[1]=
$$1 - \frac{x^2}{2}$$

In[2]:=
```
Plot[{f[x],approx2[x]},{x,-4,4},
PlotStyle->{
{Blue,Thickness[0.02]},
{Red,Thickness[0.01]}},
AxesLabel->{"x",""},
PlotRange->{-2,2}];
```

> Copy, paste, edit, and run in an effort to see what happens as you use more and more terms of the expansion of $f[x]$ in powers of x. Animate your plots. Describe what you see.

G.1.b) Go with $f[x] = e^x$. Here is a plot $f[x]$ and the polynomial made from using the terms through the x^2 term of its expansion in powers of x:

In[3]:=
```
Clear[f,approx2,x]; f[x_] = E^x;
approx2[x_] = Normal[Series[f[x],{x,0,2}]]
```

Out[3]=
$$1 + x + \frac{x^2}{2}$$

In[4]:=
```
Plot[{f[x],approx2[x]},{x,-3,3},
PlotStyle->{
{Blue,Thickness[0.02]},
{Red,Thickness[0.01]}},
AxesLabel->{"x",""},
PlotRange->{-1,E^3}];
```

> Copy, paste, edit, and run in an effort to see what happens as you use more and more terms of the expansion of $f[x]$ in powers of x. Animate your plots. Describe what you see.

Lesson 4.02 Expansions in Powers of x

G.1.c.i) Go with $f[x] = 8/(1+x^2)$. Here is a plot $f[x]$ and the polynomial made from using the terms through the x^2 term of its expansion in powers of x:

In[5]:=
```
Clear[f,approx2,x]; f[x_] = 8/(1 + x^2);
approx2[x_] = Normal[Series[f[x],{x,0,2}]]
```

Out[5]=
$$8 - 8x^2$$

In[6]:=
```
Plot[{f[x],approx2[x]},{x,-1.3,1.3},
PlotStyle->{
{Blue,Thickness[0.02]},
{Red,Thickness[0.01]}},
AxesLabel->{"x",""},
PlotRange->{-1,9}];
```

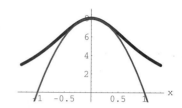

> Copy, paste, edit, and run in an effort to see what happens as you use more and more terms of the expansion of $f[x]$ in powers of x. Animate your plots. Describe what you see.

G.1.c.ii) In part G.1.c.i), do you think that you can break the barriers at $x = 1$ and $x = -1$?

G.1.d) Go with $f[x] = 10\, x^2\, e^{-x^2}$. Here is a plot $f[x]$ and the polynomial made from using the terms through the x^2 term of its expansion in powers of x:

In[7]:=
```
Clear[f,approx2,x]; f[x_] = 10 x^2 E^(-x^2);
approx2[x_] = Normal[Series[f[x],{x,0,2}]]
```

Out[7]=
$$10 x^2$$

In[8]:=
```
Plot[{f[x],approx2[x]},{x,-2.5,2.5},
PlotStyle->{{Blue,Thickness[0.02]},
{Red,Thickness[0.01]}},
AxesLabel->{"x",""},
PlotRange->{-1,9}];
```

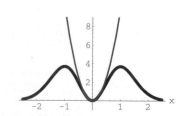

> Copy, paste, edit, and run in an effort to see what happens as you use more and more terms of the expansion of $f[x]$ in powers of x. Animate your plots. Describe what you see.

G.1.e) Write a few words on how you think the plot of the sum of the first few terms of the expansion of a given function is related to the plot of the given function. What is the effect of using more and more terms?

■ G.2) Writing down expansions★

Use the basic expansions

→ $1/(1-x)$ in powers of x:
$$1 + x + x^2 + x^3 + x^4 + \cdots + x^k + \cdots$$

→ e^x in powers of x:
$$1 + x + \frac{x^2}{2} + \frac{x^3}{3!} + \cdots + \frac{x^k}{k!} + \cdots$$

→ $\sin[x]$ in powers of x:
$$x - \frac{x^3}{3!} + \frac{x^5}{5!} + \cdots + \frac{(-1)^k x^{2k+1}}{(2k+1)!} + \cdots$$

→ $\cos[x]$ in powers of x:
$$1 - \frac{x^2}{2!} + \frac{x^4}{4!} + \cdots + \frac{(-1)^k x^{2k}}{(2k)!} + \cdots$$

to arrive at the expansions in powers of x of the following functions $f[x]$.

G.2.a) $f[x] = e^{-x^2}$

G.2.b) $f[x] = \sin[2x^2]$

G.2.c) $f[x] = \cos[2x^2]$

G.2.d.i) $f[x] = \dfrac{1}{1+x}$

G.2.d.ii) $f[x] = \dfrac{1}{1+x^8}$

Lesson 4.02 Expansions in Powers of x

G.2.d.iii) $f[x] = \dfrac{x}{1+x^8}$

G.2.e.i) $f[x] = \log[1+x]$

G.2.e.ii) $f[x] = \log[1+x^5]$

■ G.3) Trick questions*

G.3.a) Give the expansion in powers of x of the function $f[x] = 1 + 2x + 3x^2 + 4x^3$.

G.3.b) What do you get when you differentiate the expansion of $\sin[x]$ in powers of x?
What do you get when you differentiate the expansion of $\cos[x]$ in powers of x?
What do you get when you differentiate the expansion of e^x in powers of x?

■ G.4) Serious approximations*

G.4.a) Find a polynomial $p[x]$ that you can use to calculate $\cos[x]$ within an error of no more than 10^{-6} for all the x's with $-\pi/2 \leq x \leq \pi/2$.

G.4.b.i) Find a polynomial $p[x]$ that you can use to calculate $6\arctan[x]$ within an error of no more than 10^{-5} for all the x's with $-1/\sqrt{3} \leq x \leq 1/\sqrt{3}$.

G.4.b.ii) Look at this info:

In[9]:=
```
Tan[Pi/6]
```
Out[9]=
$$\dfrac{1}{\texttt{Sqrt[3]}}$$

This tells you that $6\arctan[1/\sqrt{3}\,] = \pi$.

In[10]:=
```
N[Pi,10]
```

Out[10]=
3.141592654

Take your answer, $p[x]$, to part G.4.b.i) and evaluate $N[p[1/\sqrt{3}\,], 10]$. In so doing, you should pick up π to within 10^{-5}.

> Why? What would you do to calculate π with more accuracy?

Historical Marker: When they tackled the problem of finding good approximations of π in the late 1600s, the English astronomer Halley (of comet fame) and his student Sharp did by hand essentially what you are doing here—thereby earning for themselves a niche in the history of π.

■ G.5) Circles

This problem appears only in the electronic version.

■ G.6) Turning the tables*

G.6.a)
> Come up with a formula for the function $f[x]$ whose expansion in powers of x is
> $$1 + x + \frac{x^2}{2!} + \frac{x^3}{3!} + \frac{x^4}{4!} + \cdots + \frac{x^n}{n!} + \cdots$$

G.6.b)
> Come up with a formula for the function $f[x]$ whose expansion in powers of x is
> $$1 + x^3 + \frac{x^6}{2!} + \frac{x^9}{3!} + \frac{x^{12}}{4!} + \cdots + \frac{x^{3n}}{n!} + \cdots$$

G.6.c)
> Come up with a formula for the function $f[x]$ whose expansion in powers of x is
> $$\frac{x^2}{2!} + \frac{x^3}{3!} + \frac{x^4}{4!} + \cdots + \frac{x^n}{n!} + \cdots$$

G.6.d)
> Come up with a formula for the function $f[x]$ whose expansion in powers of x is
> $$1 + x + x^2 + x^3 + x^4 + \cdots + x^n + \cdots$$

Lesson 4.02 Expansions in Powers of x

G.6.e) Come up with a formula for the function $f[x]$ whose expansion in powers of x is
$$1 + x^2 + x^4 + x^6 + x^8 + \cdots + x^{2n} + \cdots$$

G.6.f) Come up with a formula for the function $f[x]$ whose expansion in powers of x is
$$1 - \frac{x^2}{2^2\, 2!} + \frac{x^4}{2^4\, 4!} - \frac{x^6}{2^6\, 6!} + \cdots + \frac{(-1)^n\, x^{2n}}{2^n\, (2n)!} + \cdots$$

G.6.g) You can come up with the function $f[x]$ whose expansion in powers of x is
$$x + 2x^2 + 3x^3 + 4x^4 + \cdots + k\, x^k + \cdots$$
as follows: You know the expansion $1/(1-x)$ is
$$1 + x + x^2 + x^3 + x^4 + \cdots + x^k + \cdots$$
So the expansion of $D[1/(1-x), x] = 1/(1-x)^2$ is
$$1 + 2x + 3x^2 + 4x^3 + 5x^4 + \cdots + k\, x^{k-1} + \cdots$$
Consequently,
$$x + 2x^2 + 3x^3 + 4x^4 + \cdots + k\, x^k + \cdots$$
is the expansion of $f[x] = x/(1-x)^2$. Check:

In[11]:=
```
Clear[f,x]; f[x_] = x/(1 - x)^2;
Normal[Series[f[x],{x,0,9}]]
```
Out[11]=
```
          2     3     4     5     6     7     8     9
   x + 2 x + 3 x + 4 x + 5 x + 6 x + 7 x + 8 x + 9 x
```

No problem-o.

Your mission, should you choose to accept it, is to come up with a formula for the function $f[x]$ whose expansion in powers of x is
$$x + 2^2 x^2 + 3^2 x^3 + 4^2 x^4 + \cdots + k^2 x^k + \cdots$$

■ G.7) Integrating expansions

G.7.a) Start with the expansion of $\sin[t]$ in powers of t:
$$t - \frac{t^3}{3!} + \frac{t^5}{5!} + \cdots + \frac{(-1)^k\, t^{2k+1}}{(2k+1)!} + \cdots$$

> Use the fact that
> $$\int_0^x \sin[t]\, dt = \cos[x] - 1$$
> to arrive at the expansion of $\cos[x]$ in powers of x.

G.7.b) Isaac Newton worked on this very problem to illustrate the power of the fundamental formula of calculus.

Look at the expansion of $1/\sqrt{1-t^2}$ in powers of t through the t^{10} term:

In[12]:=
```
Clear[t]; Normal[Series[1/Sqrt[1 - t^2],{t,0,10}]]
```

Out[12]=
$$1 + \frac{t^2}{2} + \frac{3t^4}{8} + \frac{5t^6}{16} + \frac{35t^8}{128} + \frac{63t^{10}}{256}$$

> Recall that $\arcsin[x] = \int_0^x 1/\sqrt{1-t^2}\, dt$ and integrate to get the expansion of $\arcsin[x]$ in powers of x through the x^{11} term.
>
> Check yourself with *Mathematica*.

LESSON 4.03

Using Expansions

Basics

■ **B.1) Expansions in powers of $(x - b)$**

B.1.a) When you want to study behavior near a point $x = b$ other than $x = 0$, you can use expansions in powers of $(x - b)$.

> What is the expansion of a function $f[x]$ in powers of $(x - b)$?

Answer: Given a function $f[x]$, the expansion of $f[x]$ in powers of $(x - b)$ is
$$a[0] + a[1](x - b) + a[2](x - b)^2$$
$$+ \cdots + a[k](x - b)^k + a[k+1](x - b)^{k+1} + \cdots + \cdots$$

where the numbers $a[0], a[1], a[2], \ldots, a[k], a[k+1], \ldots$ are set so that for every positive integer m, the function $f[x]$ and the polynomial

$$a[0] + a[1](x - b) + a[2](x - b)^2 + a[3](x - b)^3 + \cdots + a[m](x - b)^m$$

have order of contact m at $x = b$.

B.1.b.i) > Come up with the expansion of $f[x] = 1/(1 + x)$ in powers of $(x - 2)$.

Answer: Look at:

Lesson 4.03 Using Expansions

In[1]:=
```
Clear[f,x,expan3]; f[x_] = 1/( 1 + x);
expan3[x_] = Normal[Series[f[x],{x,2,3}]]
```

Out[1]=

$$\frac{1}{3} - \frac{-2 + x}{9} + \frac{(-2 + x)^2}{27} - \frac{(-2 + x)^3}{81}$$

This polynomial has order of contact 3 with $f[x]$ at $x = 2$. You can check that:

In[2]:=
```
Table[{D[f[x],{x,k}],D[expan3[x],{x,k}]},{k,0,4}]/.x->2
```

Out[2]=

$$\{\{\frac{1}{3}, \frac{1}{3}\}, \{-(\frac{1}{9}), -(\frac{1}{9})\}, \{\frac{2}{27}, \frac{2}{27}\}, \{-(\frac{2}{27}), -(\frac{2}{27})\}, \{\frac{8}{81}, 0\}\}$$

Look at this:

In[3]:=
```
Normal[Series[f[x],{x,2,6}]]
```

Out[3]=

$$\frac{1}{3} + \frac{2 - x}{9} + \frac{(-2 + x)^2}{27} - \frac{(-2 + x)^3}{81} + \frac{(-2 + x)^4}{243} - \frac{(-2 + x)^5}{729} + \frac{(-2 + x)^6}{2187}$$

This polynomial has order of contact 6 with $f[x]$ at $x = 2$. The denominators are powers of 3, and the signs alternate.

The expansion of $f[x] = 1/(1 + x)$ in powers of $(x - 2)$ is

$$\frac{1}{3} - \frac{x-2}{3^2} + \frac{(x-2)^2}{3^3} - \frac{(x-2)^3}{3^4} + \cdots + \frac{(-1)^k (x-2)^k}{3^{k+1}} + \cdots$$

This isn't worth memorizing.

B.1.b.ii) Come up with the expansion of $f[x] = \cos[x]$ in powers of $(x - \pi/2)$.

Answer: Look at:

In[4]:=
```
Clear[f,x,expan3]; f[x_] = Cos[x];
expan3[x_] = Normal[Series[f[x],{x,Pi/2,3}]]
```

Out[4]=

$$\frac{Pi}{2} - x + \frac{(\frac{-Pi}{2} + x)^3}{6}$$

This polynomial has order of contact 3 with $f[x]$ at $x = \pi/2$. You can check it:

In[5]:=
```
Table[{D[f[x],{x,k}],D[expan3[x],{x,k}]},{k,0,3}]/.x->Pi/2
```

Out[5]=
{{0, 0}, {-1, -1}, {0, 0}, {1, 1}}

Look at:

In[6]:=
```
Normal[Series[f[x],{x,Pi/2,12}]]
```

Out[6]=
$$\frac{Pi}{2} - x + \frac{(\frac{-Pi}{2} + x)^3}{6} - \frac{(\frac{-Pi}{2} + x)^5}{120} + \frac{(\frac{-Pi}{2} + x)^7}{5040} - \frac{(\frac{-Pi}{2} + x)^9}{362880} + \frac{(\frac{-Pi}{2} + x)^{11}}{39916800}$$

This polynomial has order of contact 12 with $f[x]$ at $x = \pi/2$.

The expansion of $f[x] = \cos[x]$ in powers of $(x - \pi/2)$ is

$$-\left(x - \frac{\pi}{2}\right) + \frac{\left(x - \frac{\pi}{2}\right)^3}{3!} - \frac{\left(x - \frac{\pi}{2}\right)^5}{5!} + \cdots + \frac{(-1)^{k+1}\left(x - \frac{\pi}{2}\right)^{2k+1}}{(2k+1)!} + \cdots$$

B.1.c) When you expand in powers of $(x - 0)$, you get the expansion in powers of x:

In[7]:=
```
b = 0; Clear[f,approx,x]; f[x_] = E^x;
approx[x_] = Normal[Series[f[x],{x,b,6}]]
```

Out[7]=
$$1 + x + \frac{x^2}{2} + \frac{x^3}{6} + \frac{x^4}{24} + \frac{x^5}{120} + \frac{x^6}{720}$$

In[8]:=
```
Plot[{f[x],approx[x]},
{x,b - 1,b + 1},PlotStyle->
{{Thickness[0.02],Blue},{Thickness[0.01],Red}},
AxesLabel->{"x",""},PlotLabel->
"Powers of (x - 0) centered at x = 0",
Epilog->{{Blue,PointSize[0.04],Point[{b,0}]},
Text["expansion point",{b,0},{0,-2}]}];
```

Here's what happens when you do the same thing, except this time expand in powers of $(x - b)$ for $b = 1$:

In[9]:=
```
b = 1; Clear[f,approx,x]; f[x_] = E^x;
approx[x_] = Normal[Series[f[x],{x,b,6}]]
```

Out[9]=
$$E + E\,(-1 + x) + \frac{E\,(-1 + x)^2}{2} + \frac{E\,(-1 + x)^3}{6} + \frac{E\,(-1 + x)^4}{24} + \frac{E\,(-1 + x)^5}{120} + \frac{E\,(-1 + x)^6}{720}$$

```
In[10]:=
    Plot[{f[x],approx[x]},
    {x,b - 1,b + 1},PlotStyle->
    {{Thickness[0.02],Blue},{Thickness[0.01],Red}},
    AxesLabel->{"x",""},PlotLabel->
    "Powers of (x - 1) centered at x = 1",
    PlotRange->All,Epilog->
    {{Blue,PointSize[0.04],Point[{b,1}]},
    Text["expansion point",{b,1},{0,-2}]}];
```

Here's what happens when you do the same thing, except this time expand in powers of $(x - b)$ for $b = 0.8$:

```
In[11]:=
    b = 0.8; Clear[f,approx,x]; f[x_] = E^x;
    approx[x_] = Normal[Series[f[x],{x,b,6}]]
Out[11]=
                                                 2                 3
    2.22554 + 2.22554 (-0.8 + x) + 1.11277 (-0.8 + x)  + 0.370923 (-0.8 + x)  +
                        4                      5                      6
    0.0927309 (-0.8 + x)  + 0.0185462 (-0.8 + x)  + 0.00309103 (-0.8 + x)
```

```
In[12]:=
    Plot[{f[x],approx[x]},
    {x,b - 1,b + 1},PlotStyle->
    {{Thickness[0.02],Blue},{Thickness[0.01],Red}},
    AxesLabel->{"x",""},PlotLabel->
    "Powers of (x - 0.8) centered at x = 0.8",
    PlotRange->All,Epilog->
    {{Blue,PointSize[0.04],Point[{b,1}]},
    Text["expansion point",{b,1},{0,-2}]}];
```

What's the message?

Answer: Expansions in powers of $(x - b)$ give you the same high quality approximations near $x = b$ that expansions in powers of x give you near $x = 0$. The reason: High order of contact at $x = b$.

■ B.2) Tangent lines and Newton's method.

B.2.a) Look at these:

```
In[13]:=
    b = 1; Clear[f,x,expan1]; f[x_] = E^x - 1;
    expan1[x_] = Normal[Series[f[x],{x,b,1}]]
Out[13]=
    -1 + E + E (-1 + x)
```

In[14]:=
```
Plot[{f[x],expan1[x]},
 {x,b - 1,b + 1},
 PlotStyle->
 {{Blue,Thickness[0.01]},
 {Red,Thickness[0.01]}},
 AxesLabel->{"x",""}];
```

In[15]:=
```
b = 0.6; Clear[f,x,expan1]; f[x_] = 1/(1 + x^2) - x;
expan1[x_] = Normal[Series[f[x],{x,b,1}]]
```

Out[15]=
0.135294 - 1.64879 (-0.6 + x)

In[16]:=
```
Plot[{f[x],expan1[x]},
 {x,b - 1,b + 1},
 PlotStyle->
 {{Blue,Thickness[0.01]},
 {Red,Thickness[0.01]}},
 AxesLabel->{"x",""}];
```

In[17]:=
```
b = Pi/4; Clear[f,x,expan1]; f[x_] = Sin[x];
expan1[x_] = Normal[Series[f[x],{x,b,1}]]
```

Out[17]=

$$\frac{1}{\text{Sqrt}[2]} + \frac{\frac{-\text{Pi}}{4} + x}{\text{Sqrt}[2]}$$

In[18]:=
```
Plot[{f[x],expan1[x]},
 {x,b - 1,b + 1},
 PlotStyle->
 {{Blue,Thickness[0.01]},
 {Red,Thickness[0.01]}},
 AxesLabel->{"x",""}];
```

> Explain what you see.

Answer: You guessed right. When you expand a function $f[x]$ in powers of $(x - b)$ through the $(x - b)^1$ term, then you get the line function that is tangent to the plot of $f[x]$ at the point $\{b, f[b]\}$.

B.2.b) Look at this plot of $f[x] = x - \cos[x]$.

Lesson 4.03 Using Expansions

In[19]:=
```
Clear[f,x]
f[x_] = x - Cos[x];
plot = Plot[f[x],
  {x,-0.5,1.5},PlotStyle->
  {{Blue,Thickness[0.01]}},
  AxesLabel->{"x","y"}];
```

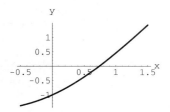

The plot tells you that the number x^* for which $f[x^*] = 0$ is reasonably approximated by $x^* = 0.7$. Unfortunately, *Mathematica* cannot deliver this number from the usual Solve instruction:

In[20]:=
```
Solve[f[x] == 0,x]
```
Solve::ifun:
   Warning: Inverse functions are being used
     by Solve, so some solutions may not be
     found.
Solve::tdep:
   The equations appear to involve
     transcendental functions of the
     variables in an essentially
     non-algebraic way.

Out[20]=
```
Solve[x - Cos[x] == 0, x]
```

Reason: cos[x] is a transcendental function that is highly resistant to algebra.

> What can you do to coax a reasonable estimate of the number x^* with $f[x^*] = 0$ from the machine?

Answer: Take 0.7 as your rough approximation and and look at this plot of $f[x]$ together with the tangent line through $\{0.7, f[0.7]\}$:

In[21]:=
```
b = 0.7; Clear[f,x,expan1]; f[x_] = x - Cos[x];
expan1[x_] = Normal[Series[f[x],{x,b,1}]]
```

Out[21]=
```
-0.0648422 + 1.64422 (-0.7 + x)
```

In[22]:=
```
plot = Plot[{f[x],expan1[x]},
  {x,b - 0.5,b + 0.5},
  PlotStyle->{{Blue,Thickness[0.01]},
  {Red,Thickness[0.01]}},
  AxesLabel->{"x","y"}];
```

The plot of $f[x]$ and the plot of the tangent line cross the line $y = 0$ at nearly the same point. And the tangent line function is not resistant to algebra:

In[23]:=
```
Solve[expan1[x] == 0,x]
```
Out[23]=
```
{{x -> 0.739436}}
```

Try it out:

In[24]:=
```
f[0.739436]
```
Out[24]=
```
0.00058726
```

Pretty good. To get a better estimate, do it again.

In[25]:=
```
b = 0.739436;
expan1[x_] = Normal[Series[f[x],{x,b,1}]];
Solve[expan1[x] == 0,x]
```
Out[25]=
```
{{x -> 0.739085}}
```

Try it out:

In[26]:=
```
f[0.739085]
```
Out[26]=
```
-2.2295 10^-7
```

Very close to 0. Now see what *Mathematica*'s FindRoot instruction does:

In[27]:=
```
FindRoot[f[x] == 0,{x,0.7}]
```
Out[27]=
```
{x -> 0.739085}
```

Hey, it got the same result as above.

The reason: It used the same method as used above. Folks like to call this method Newton's Method. This is the same Newton who, along with Leibniz, invented calculus. Old Newton had a lot of brains.

B.2.c) How about a systematic way of doing Newton's Method for going after a good approximation to a true solution x^* of $f[x^*] = 0$ for any old function $f[x]$?

Answer: Start with a reasonable guess $a[1]$ for x.

The tangent line through $\{a[1], f[a[1]]\}$ crosses the x-axis at a number $a[2]$. The equation of this tangent line is

$$y - f[a[1]] = f'[a[1]] \, (x - a[1]).$$

Lesson 4.03 Using Expansions

It crosses the x-axis at a point $a[2]$ satisfying
$$0 - f[a[1]] = f'[a[1]]\,(a[2] - a[1]).$$
Solve this for $a[2]$, getting
$$a[2] = a[1] - \frac{f[a[1]]}{f'[a[1]]}.$$
Iterate this idea via the update formula
$$a[n] = a[n-1] - \frac{f[a[n-1]]}{f'[a[n-1]]}$$
which you get by making the replacements $2 \to n$ and $1 \to (n-1)$ in the formula expressing $a[2]$ in terms of $a[1]$.

Turn it over to *Mathematica* and run it on a sample problem; finding a good approximate solution of $e^x - x - 2 = 0$:

In[28]:=
```
Clear[f,x]; f[x_] = E^x - x - 2;
```

Plot:

In[29]:=
```
fplot = Plot[f[x],{x,0,2},
  PlotStyle->
  {{Thickness[0.01],Blue}},
  AxesLabel->{"x","f[x]"}];
```

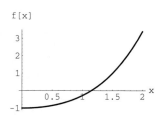

A reasonable choice for $a[1]$ is:

In[30]:=
```
Clear[a]; a[1] = 1;
```

The update formula is:

In[31]:=
```
Clear[n]; a[n_] := a[n] = a[n-1] - f[a[n-1]]/f'[a[n-1]]
```

Take a look at what comes out:

In[32]:=
```
Table[N[a[n]],{n,1,5}]
```

Out[32]=
```
{1., 1.16395, 1.14642, 1.14619, 1.14619}
```

The approximations stabilize at 1.14619. Check it out:

In[33]:=
```
f[1.14619]
```

Out[33]=
$$-6.91206\;10^{-6}$$

Darn close to 0. See what FindRoot does starting at 1:

In[34]:=
  FindRoot[f[x] == 0,{x,1}]

Out[34]=
  {x -> 1.14619}

This was no accident. See it in plot form:

In[35]:=
  Show[fplot,
   Graphics[
    {Black,PointSize[0.03],
    Point[{1.14619,f[1.14619]}]}]];

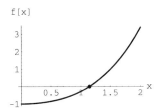

Nailed it.

B.2.d) The Newton iteration formula for finding an approximate solution of $f[x] = 0$ is
$$a[n] = a[n-1] - \frac{f[a[n-1]]}{f'[a[n-1]]}.$$

If the solution x_{root} has the property that $f'[x_{\text{root}}] = 0$, then this can make Newton's method less than reliable because the iteration will have the tendency to divide by very small numbers. Look at:

In[36]:=
  FindRoot[x == 0,{x,0.2}]

Out[36]=
  {x -> 0.}

Good.

In[37]:=
  FindRoot[x^2 == 0,{x,0.2}]

Out[37]=
  {x -> 0.000390625}

Not bad.

In[38]:=
  FindRoot[x^4 == 0,{x,0.2}]

Out[38]=
  {x -> 0.0200226}

Suspicious, but at the true solution the derivative is 0.

In[39]:=
  FindRoot[x^8 == 0,{x,0.2}]

Out[39]=
  {x -> 0.153125}

Awful. The trouble you are seeing here in not with mathematics or with *Mathematica*. What you are seeing is Newton's method breaking down because it was applied under unfavorable conditions.

■ B.3) The *Mathematica* Series instruction

B.3.a.i) By now, you are used to the instruction:

In[40]:=
```
Clear[x]; Normal[Series[E^x,{x,0,6}]]
```
Out[40]=
$$1 + x + \frac{x^2}{2} + \frac{x^3}{6} + \frac{x^4}{24} + \frac{x^5}{120} + \frac{x^6}{720}$$

This is the sixth degree polynomial that has order of contact 6 with $f[x] = e^x$ at $x = 0$. Now look at:

In[41]:=
```
Series[E^x,{x,0,6}]
```
Out[41]=
$$1 + x + \frac{x^2}{2} + \frac{x^3}{6} + \frac{x^4}{24} + \frac{x^5}{120} + \frac{x^6}{720} + O[x]^7$$

That O on the left is not a zero; it is a capital Oh.

> What does that notation mean?

Answer: Try it again:

In[42]:=
```
Table[Series[E^x,{x,0,k}],{k,0,5}]//ColumnForm
```
Out[42]=
$$1 + O[x]$$
$$1 + x + O[x]^2$$
$$1 + x + \frac{x^2}{2} + O[x]^3$$
$$1 + x + \frac{x^2}{2} + \frac{x^3}{6} + O[x]^4$$
$$1 + x + \frac{x^2}{2} + \frac{x^3}{6} + \frac{x^4}{24} + O[x]^5$$
$$1 + x + \frac{x^2}{2} + \frac{x^3}{6} + \frac{x^4}{24} + \frac{x^5}{120} + O[x]^6$$

The big $O[x]^k$ notation just stands for the higher degree terms of the expansion in powers of x that are not present in the output. If you don't want the $O[x]^k$ term, then you can get rid of it by hanging Normal[ ] around it:

In[43]:=
```
{Series[E^x,{x,0,4}], Normal[Series[E^x,{x,0,4}]]}
```

Out[43]=
$$\left\{1 + x + \frac{x^2}{2} + \frac{x^3}{6} + \frac{x^4}{24} + O[x]^5,\ 1 + x + \frac{x^2}{2} + \frac{x^3}{6} + \frac{x^4}{24}\right\}$$

B.3.a.ii) Look at this:

In[44]:=
```
Clear[x]; Normal[Series[E^x,{x,2,6}]]
```

Out[44]=
$$E^2 + E^2(-2+x) + \frac{E^2(-2+x)^2}{2} + \frac{E^2(-2+x)^3}{6} + \frac{E^2(-2+x)^4}{24} + \frac{E^2(-2+x)^5}{120} + \frac{E^2(-2+x)^6}{720}$$

This is the sixth degree polynomial that has order of contact 6 with $f[x] = e^x$ at $x = 2$. Now look at:

In[45]:=
```
Series[E^x,{x,2,6}]
```

Out[45]=
$$E^2 + E^2(-2+x) + \frac{E^2(-2+x)^2}{2} + \frac{E^2(-2+x)^3}{6} + \frac{E^2(-2+x)^4}{24} + \frac{E^2(-2+x)^5}{120} + \frac{E^2(-2+x)^6}{720} + O[-2+x]^7$$

That O on the left is not a zero; it is a capital Oh.

What does that notation mean?

Answer: Try it again:

In[46]:=
```
Table[Series[E^x,{x,2,k}],{k,0,4}]//ColumnForm
```

Out[46]=
$$E^2 + O[-2+x]$$
$$E^2 + E^2(-2+x) + O[-2+x]^2$$

$$E^2 + E^2(-2+x) + \frac{E^2(-2+x)^2}{2} + O[-2+x]^3$$

$$E^2 + E^2(-2+x) + \frac{E^2(-2+x)^2}{2} + \frac{E^2(-2+x)^3}{6} + O[-2+x]^4$$

$$E^2 + E^2(-2+x) + \frac{E^2(-2+x)^2}{2} + \frac{E^2(-2+x)^3}{6} + \frac{E^2(-2+x)^4}{24} + O[-2+x]^5$$

The big $O[x-2]^k$ notation just stands for the higher degree terms of the expansion in powers of $(x-2)$ that are not present in the output. If you don't want the $O[x-2]^k$ term, then you can get rid of it by hanging Normal[ ] around it:

In[47]:=
```
Series[E^x,{x,2,4}]
Normal[Series[E^x,{x,2,4}]]
```

Out[47]=

$$E^2 + E^2(-2+x) + \frac{E^2(-2+x)^2}{2} + \frac{E^2(-2+x)^3}{6} + \frac{E^2(-2+x)^4}{24} + O[-2+x]^5$$

Out[47]=

$$E^2 + E^2(-2+x) + \frac{E^2(-2+x)^2}{2} + \frac{E^2(-2+x)^3}{6} + \frac{E^2(-2+x)^4}{24}$$

B.3.b) Explain the output from:

In[48]:=
```
Clear[x]; Series[E^x Cos[Pi x],{x,2,4}]
```

Out[48]=

$$E^2 + E^2(-2+x) + \left(\frac{E^2}{2} - \frac{E^2 Pi^2}{2}\right)(-2+x)^2 + \left(\frac{E^2}{6} - \frac{E^2 Pi^2}{2}\right)(-2+x)^3 + \left(\frac{E^2}{24} - \frac{E^2 Pi^2}{4} + \frac{E^2 Pi^4}{24}\right)(-2+x)^4 + O[-2+x]^5$$

Answer: This must come from the expansion of e^x in powers of $(x-2)$ multiplied by the expansion of $\cos[\pi x]$ in powers of $(x-2)$. Check this out:

In[49]:=
```
Eexpansion = Series[E^x,{x,2,4}]
```

Out[49]=

$$E^2 + E^2(-2+x) + \frac{E^2(-2+x)^2}{2} + \frac{E^2(-2+x)^3}{6} + \frac{E^2(-2+x)^4}{24} + O[-2+x]^5$$

In[50]:=
```
Cosexpansion = Series[Cos[Pi x],{x,2,4}]
```

Basics (B.3)

Out[50]=
$$1 - \frac{Pi^2 (-2+x)^2}{2} + \frac{Pi^4 (-2+x)^4}{24} + O[-2+x]^5$$

Their product is:

In[51]:=
```
Eexpansion Cosexpansion
```

Out[51]=
$$E^2 + E^2 (-2+x) + \left(\frac{E^2}{2} - \frac{E^2 Pi^2}{2}\right)(-2+x)^2 + \left(\frac{E^2}{6} - \frac{E^2 Pi^2}{2}\right)(-2+x)^3 +$$
$$\left(\frac{E^2}{24} - \frac{E^2 Pi^2}{4} + \frac{E^2 Pi^4}{24}\right)(-2+x)^4 + O[-2+x]^5$$

Compare with:

In[52]:=
```
Series[E^x Cos[Pi x],{x,2,4}]
```

Out[52]=
$$E^2 + E^2 (-2+x) + \left(\frac{E^2}{2} - \frac{E^2 Pi^2}{2}\right)(-2+x)^2 + \left(\frac{E^2}{6} - \frac{E^2 Pi^2}{2}\right)(-2+x)^3 +$$
$$\left(\frac{E^2}{24} - \frac{E^2 Pi^2}{4} + \frac{E^2 Pi^4}{24}\right)(-2+x)^4 + O[-2+x]^5$$

Beware! This is not the same as:

In[53]:=
```
wrong = Expand[Normal[Eexpansion] Normal[Cosexpansion]]
```

Out[53]=
$$\frac{E^2}{3} - \frac{2 E^2 Pi^2}{3} + \frac{2 E^2 Pi^4}{9} - \frac{2 E^2 Pi^2}{3} \cdot x + \frac{4 E^2 Pi^2 x}{3} - \frac{2 E^2 Pi^4 x}{3} + \frac{E^2 x^2}{2} -$$
$$\frac{11 E^2 Pi^2 x^2}{6} + \frac{10 E^2 Pi^4 x^2}{9} - \frac{E^2 x^3}{6} + \frac{3 E^2 Pi^2 x^3}{2} - \frac{11 E^2 Pi^4 x^3}{9} + \frac{E^2 x^4}{24} -$$
$$\frac{2 E^2 Pi^2 x^4}{3} + \frac{7 E^2 Pi^4 x^4}{8} + \frac{E^2 Pi^2 x^5}{6} - \frac{29 E^2 Pi^4 x^5}{72} - \frac{E^2 Pi^6 x^5}{48} +$$
$$\frac{17 E^2 Pi^4 x^6}{144} - \frac{E^2 Pi^4 x^7}{48} + \frac{E^2 Pi^4 x^8}{576}$$

In[54]:=
```
correct = Expand[Normal[Eexpansion Cosexpansion]]
```

Lesson 4.03 Using Expansions

Out[54]=

$$\frac{E^2}{3} - 2E^2 Pi^2 + \frac{2E^2 Pi^4}{3} - \frac{2E^2 x}{3} + 4E^2 Pi^2 x - \frac{4E^2 Pi^4 x}{3} + \frac{E^2 x^2}{2} - \frac{7E^2 Pi^2 x^2}{2} + E^2 Pi^4 x^2 - \frac{2E^2 x^3}{6} + \frac{3E^2 Pi^2 x^3}{2} - \frac{E^2 Pi^4 x^3}{3} + \frac{E^2 x^4}{24} - \frac{E^2 Pi^2 x^4}{4} + \frac{E^2 Pi^4 x^4}{24}$$

The wrong output above is accurate through the fourth degree terms only. If you use the Series instruction without hanging Normal[ ] around it, then *Mathematica* gives you only the terms that are accurate.

Nice going *Mathematica*!

B.3.c) Explain the output from:

In[55]:=
```
Clear[x]; Series[(E^x)/Cos[x],{x,Pi/4,3}]
```

Out[55]=

$$\text{Sqrt}[2]\, E^{Pi/4} + 2\,\text{Sqrt}[2]\, E^{Pi/4}\left(\frac{-Pi}{4} + x\right) + 3\,\text{Sqrt}[2]\, E^{Pi/4}\left(\frac{-Pi}{4} + x\right)^2 + 4\,\text{Sqrt}[2]\, E^{Pi/4}\left(\frac{-Pi}{4} + x\right)^3 + O\left[\frac{-Pi}{4} + x\right]^4$$

Answer: This must come from the expansion of e^x in powers of $(x - \pi/4)$ divided by the expansion of $\cos[x]$ in powers of $(x - \pi/4)$. Check this out:

In[56]:=
```
Eexpansion = Series[E^x,{x,Pi/4,3}]
```

Out[56]=

$$E^{Pi/4} + E^{Pi/4}\left(\frac{-Pi}{4} + x\right) + \frac{E^{Pi/4}\left(\frac{-Pi}{4} + x\right)^2}{2} + \frac{E^{Pi/4}\left(\frac{-Pi}{4} + x\right)^3}{6} + O\left[\frac{-Pi}{4} + x\right]^4$$

In[57]:=
```
Cosexpansion = Series[Cos[x],{x,Pi/4,3}]
```

Out[57]=

$$\frac{1}{\text{Sqrt}[2]} - \frac{\frac{-Pi}{4} + x}{\text{Sqrt}[2]} - \frac{\left(\frac{-Pi}{4} + x\right)^2}{2\,\text{Sqrt}[2]} + \frac{\left(\frac{-Pi}{4} + x\right)^3}{6\,\text{Sqrt}[2]} + O\left[\frac{-Pi}{4} + x\right]^4$$

Their quotient is:

In[58]:=
```
Eexpansion/Cosexpansion
```

Out[58]=

$$\sqrt{2}\, E^{Pi/4} + 2\sqrt{2}\, E^{Pi/4}\left(\frac{-Pi}{4}+x\right) + 3\sqrt{2}\, E^{Pi/4}\left(\frac{-Pi}{4}+x\right)^2 +$$

$$4\sqrt{2}\, E^{Pi/4}\left(\frac{-Pi}{4}+x\right)^3 + O\left[\frac{-Pi}{4}+x\right]^4$$

Compare with:

In[59]:=
```
Series[(E^x)/Cos[x],{x,Pi/4,3}]
```

Out[59]=

$$\sqrt{2}\, E^{Pi/4} + 2\sqrt{2}\, E^{Pi/4}\left(\frac{-Pi}{4}+x\right) + 3\sqrt{2}\, E^{Pi/4}\left(\frac{-Pi}{4}+x\right)^2 +$$

$$4\sqrt{2}\, E^{Pi/4}\left(\frac{-Pi}{4}+x\right)^3 + O\left[\frac{-Pi}{4}+x\right]^4$$

On the mark!

But if you clip off the $O[x]^k$ terms, then you get something not very useful:

In[60]:=
```
Normal[Eexpansion]
```

Out[60]=

$$E^{Pi/4} + E^{Pi/4}\left(\frac{-Pi}{4}+x\right) + \frac{E^{Pi/4}\left(\frac{-Pi}{4}+x\right)^2}{2} + \frac{E^{Pi/4}\left(\frac{-Pi}{4}+x\right)^3}{6}$$

In[61]:=
```
Normal[Cosexpansion]
```

Out[61]=

$$\frac{1}{\sqrt{2}} - \frac{\frac{-Pi}{4}+x}{\sqrt{2}} - \frac{\left(\frac{-Pi}{4}+x\right)^2}{2^{3/2}} + \frac{\left(\frac{-Pi}{4}+x\right)^3}{3\cdot 2^{3/2}}$$

In[62]:=
```
Normal[Eexpansion]/Normal[Cosexpansion]
```

Out[62]=

$$\frac{E^{Pi/4} + E^{Pi/4}\left(\frac{-Pi}{4}+x\right) + \frac{E^{Pi/4}\left(\frac{-Pi}{4}+x\right)^2}{2} + \frac{E^{Pi/4}\left(\frac{-Pi}{4}+x\right)^3}{6}}{\frac{1}{\sqrt{2}} - \frac{\frac{-Pi}{4}+x}{\sqrt{2}} - \frac{\left(\frac{-Pi}{4}+x\right)^2}{2\sqrt{2}} + \frac{\left(\frac{-Pi}{4}+x\right)^3}{6\sqrt{2}}}$$

Lesson 4.03 Using Expansions

■ B.4) Using expansions to help to calculate limits

To get the full impact of this problem, you should be familiar with B.3 above. The need to calculate limits like

$$\lim_{x \to 0} \frac{f[x]}{g[x]}$$

comes up every so often. Usually there is no problem:

In[63]:=
```
Clear[f,g,x]; f[x_] = Cos[2 x]; g[x_] = E^(-x^2);
f[0]/g[0]
```

Out[63]=
1

No sweat. In this situation,

$$\lim_{x \to 0} \frac{f[x]}{g[x]} = 1.$$

But in other situations, there can be a complication:

In[64]:=
```
Clear[f,g,x];
f[x_] = Cos[2 x] - 1; g[x_] = E^(-x^2) - 1;
f[0]/g[0]
```
```
Power::infy:
                    1
    Infinite expression - encountered.
                    0
Infinity::indet:
    Indeterminate expression 0 ComplexInfinity
        encountered.
```
Out[64]=
    Indeterminate

The complication is:

In[65]:=
```
{f[0],g[0]}
```

Out[65]=
    {0, 0}

Tsk, tsk. $f[0] = g[0] = 0$, so you can't make sense out of $f[0]/g[0]$.

B.4.a) Use expansions to get around this obstacle.

Answer: Go with the beginnings of the expansions:

In[66]:=
```
fexpan = Series[f[x],{x,0,4}]
```

Out[66]=
$$-2x^2 + \frac{2x^4}{3} + O[x]^5$$

This has order of contact 4 with $f[x]$.

In[67]:=
```
gexpan = Series[g[x],{x,0,4}]
```

Out[67]=
$$-x^2 + \frac{x^4}{2} + O[x]^5$$

This has order of contact 4 with $g[x]$. Now divide one into the other:

In[68]:=
```
Clear[fovergapprox]
fovergapprox[x_] = Normal[fexpan/gexpan]
```

Out[68]=
$$2 + \frac{x^2}{3}$$

And plot:

In[69]:=
```
Plot[{f[x]/g[x],
 fovergapprox[x]},{x,-1,1},
 PlotStyle->{{Blue,Thickness[0.01]},
 {Red,Thickness[0.01]}},
 AxesLabel->{"x",""},PlotLabel->
 "f[x]/g[x] and a good approximation"];
```

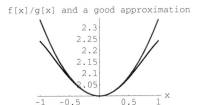

```
Power::infy:
    Infinite expression 1/0. encountered.
Infinity::indet:
    Indeterminate expression
    0. ComplexInfinity encountered.
```

Evidently, fovergapprox[x] has good contact with $f[x]/g[x]$ near $x = 0$. Based on this plot, you can say that

$$\lim_{x \to 0} \frac{f[x]}{g[x]}$$

is:

In[70]:=
```
fovergapprox[0]
```

Out[70]=
2

Check with *Mathematica*:

In[71]:=
```
Limit[f[x]/g[x],x->0]
```

Out[71]=
2

Got it. The idea of order of contact wins again.

B.4.b.i) Give a systematic procedure to use to calculate
$$\lim_{x \to 0} \frac{f[x]}{g[x]}.$$

Answer: Look again at what happened above:

In[72]:=
```
Clear[f,g,x]; f[x_] = Cos[2 x] - 1; g[x_] = E^(-x^2) - 1;
```

In[73]:=
```
fexpan = Series[f[x],{x,0,4}]
```

Out[73]=
$$-2 x^2 + \frac{2 x^4}{3} + O[x]^5$$

In[74]:=
```
gexpan = Series[g[x],{x,0,4}]
```

Out[74]=
$$-x^2 + \frac{x^4}{2} + O[x]^5$$

In[75]:=
```
fexpan/gexpan
```

Out[75]=
$$2 + \frac{x^2}{3} + O[x]^3$$

The result:
$$\lim_{x \to 0} \frac{f[x]}{g[x]} = 2.$$

In[76]:=
```
Limit[f[x]/g[x],x->0]
```

Out[76]=
2

Once you have the expansions of $f[x]$ and $g[x]$ in powers of x, you can do these in your head.

Here's how: Take the first nonzero term of the expansion of $g[x]$ in powers of x, divide it into the first nonzero term of the expansion of $f[x]$ in powers of x, and then see what happens as x closes in on 0. In the case above:

In[77]:=
fexpan

Out[77]=
$$-2x^2 + \frac{2x^4}{3} + O[x]^5$$

In[78]:=
gexpan

Out[78]=
$$-x^2 + \frac{x^4}{2} + O[x]^5$$

Divide $-x^2$ into $-2x^2$ to get
$$\lim_{x \to 0} \frac{f[x]}{g[x]} = \lim_{x \to 0} \frac{-2x^2}{x^2} = 2.$$

B.4.b.ii) Use the systematic procedure from part B.4.b.i) above to calculate
$$\lim_{x \to 0} \frac{f[x]}{g[x]} \quad \text{for} \quad f[x] = x - \sin[x] \quad \text{and} \quad g[x] = x^2.$$

Answer: Just do it!

In[79]:=
Clear[f,g,x]; f[x_] = x - Sin[x]; g[x_] = x^2;
fexpan = Series[f[x],{x,0,4}]

Out[79]=
$$\frac{x^3}{6} + O[x]^5$$

In[80]:=
gexpan = Series[g[x],{x,0,4}]

Out[80]=
$$x^2 + O[x]^5$$

Divide x^2 into $x^3/6$ to get
$$\lim_{x \to 0} \frac{f[x]}{g[x]} = \lim_{x \to 0} \frac{(x^3/6)}{x^2} = \lim_{x \to 0} \frac{x}{6} = 0.$$

Check:

In[81]:=
```
Limit[f[x]/g[x],x->0]
```
Out[81]=
0

Nailed it. And guess what? *Mathematica* probably calculated this limit just the way you did above.

B.4.b.iii) To use the systematic procedure above, do you have to check whether $f[0]$ and $g[0]$ are both 0?

Answer: NO. However, if $g[0]$ is not 0, you can calculate

$$\lim_{x \to 0} \frac{f[x]}{g[x]} = \frac{f[0]}{g[0]}$$

and save yourself the trouble of getting the expansions.

B.4.c) Adapt the systematic procedure in part B.4.b.i) above to calculate

$$\lim_{x \to 1} \frac{f[x]}{g[x]} \quad \text{for} \quad f[x] = \cos[\pi \, x] + 1 \quad \text{and} \quad g[x] = \sin[\pi \, x]^2.$$

Answer: Here you go:

In[82]:=
```
Clear[f,g,x]; f[x_] = Cos[Pi x] + 1; g[x_] = Sin[Pi x]^2;
fexpan = Series[f[x],{x,1,4}]
```
Out[82]=

$$\frac{\text{Pi}^2 \, (-1 + x)^2}{2} - \frac{\text{Pi}^4 \, (-1 + x)^4}{24} + O[-1 + x]^5$$

In[83]:=
```
gexpan = Series[g[x],{x,1,4}]
```
Out[83]=

$$\text{Pi}^2 \, (-1 + x)^2 - \frac{\text{Pi}^4 \, (-1 + x)^4}{3} + O[-1 + x]^5$$

Divide $\pi^2 (x-1)^2$ into $\pi^2 (x-1)^2/2$ to get

$$\lim_{x \to 1} \frac{f[x]}{g[x]} = \lim_{x \to 1} \frac{\pi^2 (x-1)^2 / 2}{\pi^2 (x-1)^2} = \lim_{x \to 1} \frac{1}{2} = \frac{1}{2}.$$

Check:

In[84]:=
```
Limit[f[x]/g[x],x->1]
```
Out[84]=
$$\frac{1}{2}$$

Nailed it.

■ B.5) Expansions and the complex exponential function

Some of this may be review for some of you.

B.5.a) Here is the imaginary number $i = \sqrt{-1}$:

In[85]:=
```
I;
```

Here is i^2:

In[86]:=
```
I^2
```
Out[86]=
-1

The imaginary number $i = \sqrt{-1}$ can do a lot to explain the similarity of the expansions:

$$e^x \to 1 + x + \frac{x^2}{2!} + \frac{x^3}{3!} + \frac{x^4}{4!} + \cdots + \frac{x^k}{k!} + \cdots$$

$$\sin[x] \to x - \frac{x^3}{3!} + \frac{x^5}{5!} - \cdots + \frac{(-1)^k x^{2k+1}}{(2k+1)!} + \cdots$$

$$\cos[x] \to 1 - \frac{x^2}{2!} + \frac{x^4}{4!} - \frac{x^6}{6!} + \cdots + \frac{(-1)^k x^{2k}}{(2k)!} + \cdots$$

Look at:

In[87]:=
```
Clear[x]
Normal[Series[E^x,{x,0,6}]]/.x->I x
```
Out[87]=
```
            2            4          6
           x    I  3    x    I  5   x
1 + I x - -- - - x  + -- + --- x - ---
           2    6     24   120     720
```

In[88]:=
```
Expand[Normal[Series[Cos[x],{x,0,6}]] + I Normal[Series[Sin[x],{x,0,6}]]]
```

Out[88]=
```
            2            4          6
           x    I  3    x    I  5   x
1 + I x - -- - - x  + -- + --- x - ---
           2    6     24   120     720
```

> Try to determine the relationship.

Answer: Try a few more:

In[89]:=
```
Normal[Series[E^x,{x,0,12}]]/.x->I x
```

Out[89]=

$$1 + I\,x - \frac{x^2}{2} - \frac{I\,x^3}{6} + \frac{x^4}{24} + \frac{I\,x^5}{120} - \frac{x^6}{720} - \frac{I\,x^7}{5040} + \frac{x^8}{40320} + \frac{I\,x^9}{362880} - \frac{x^{10}}{3628800} - \frac{I\,x^{11}}{39916800} + \frac{x^{12}}{479001600}$$

In[90]:=
```
Expand[Normal[Series[Cos[x],{x,0,12}]] + I Normal[Series[Sin[x],{x,0,12}]]]
```

Out[90]=

$$1 + I\,x - \frac{x^2}{2} - \frac{I\,x^3}{6} + \frac{x^4}{24} + \frac{I\,x^5}{120} - \frac{x^6}{720} - \frac{I\,x^7}{5040} + \frac{x^8}{40320} + \frac{I\,x^9}{362880} - \frac{x^{10}}{3628800} - \frac{I\,x^{11}}{39916800} + \frac{x^{12}}{479001600}$$

Again both lines came out the same. Evidently, if you take the expansion of e^x in powers of x and replace x by i times x, then you get the expansion of cos[x] plus i times the expansion of sin[x]. Another compelling reason for the definition

$$e^{ix} = \cos[x] + i\,\sin[x].$$

This relationship between the exponential function e^x and the trigonometric functions cos[x] and sin[x] is called Euler's formula.

B.5.b.i) > For a real number x, you are pretty confident about calculating e^x. But if $z = x + i\,y$ is a complex number, then how can you make sense of e^z?

Answer: Just write $z = x + i\,y$ and $e^z = e^{x+iy} = e^x e^{iy} = e^x\,(\cos[y] + i\,\sin[y])$. So,

$$e^{x+iy} = e^x\,\cos[y] + i\,e^x\,\sin[y].$$

Not much to it.

B.5.b.ii) > How do complex numbers allow you to take the natural logarithm of a negative real number?

Answer: Use the usual agreement that $y = \log[x]$ if $e^y = x$. Under this agreement, if you want $\log[-2]$, for instance, you find a complex number $y = a + i\,b$ with $e^y = -2$. The resulting complex number y will work for $\log[-2]$. So you go after a complex number $y = a + i\,b$ with

$$e^y = e^{a+i\,b} = e^a\,(\cos[b] + i\,\sin[b]) = -2.$$

The first observation is that $\sin[b]$ must be zeroed out. This gives a couple of possibilities:

In[91]:=
```
Clear[a,b]; E^a (Cos[b] + I Sin[b])/.b->0
```
Out[91]=
$$E^a$$

In[92]:=
```
E^a (Cos[b] + I Sin[b])/.b->Pi
```
Out[92]=
$$-E^a$$

The first of these cannot work because there is no real number a that makes $e^a = -2$. So go with the second. This is easy because now all you have to do is solve $-e^a = -2$ for a. This is the same as $e^a = 2$, which sets $a = \log[2]$. Try it out:

In[93]:=
```
E^a (Cos[b] + I Sin[b])/.{b->Pi,a->Log[2]}
```
Out[93]=
```
-2
```

It works! This means

$$\log[-2] = \log[2] + i\,\pi.$$

Similarly, you can see that if x is any positive real number, then $\log[-x] = \log[x] + i\,\pi$. Take a look:

In[94]:=
```
N[{Log[-2],Log[2] + I Pi},8]
```
Out[94]=
```
{0.69314718 + 3.1415927 I, 0.69314718 + 3.1415927 I}
```

In[95]:=
```
N[{Log[-E],Log[E] + I Pi},8]
```
Out[95]=
```
{1. + 3.1415927 I, 1. + 3.1415927 I}
```

Slick. Shame on that bureaucratic math teacher who once told you that you cannot take the log of a negative number. Still, this doesn't tell you how to take the log of zero. This is going to be impossible because logs of negative numbers near zero all carry the term $i\,\pi$ while the logs of positive numbers near zero do not.

Lesson 4.03 Using Expansions

B.5.b.iii) If $z = x + iy$ is a complex number, then how can you make sense of $\sin[z]$ and $\cos[z]$?

Answer: Well,
$$e^{ix} = \cos[x] + i\sin[x]$$
and
$$e^{-ix} = \cos[-x] + i\sin[-x] = \cos[x] - i\sin[x].$$

Solve for $\sin[x]$ and $\cos[x]$ in terms of e^{ix} and e^{-ix}.

In[96]:=
```
ExpandAll[Solve[{E^(I x) == Cos[x] + I Sin[x],
    E^(-I x) == Cos[x] - I Sin[x]},{Sin[x],Cos[x]}]]
```

Out[96]=
$$\{\{\text{Sin}[x] \to \frac{I}{2} E^{-I\,x} - \frac{I}{2} E^{I\,x}, \text{Cos}[x] \to \frac{E^{-I\,x}}{2} + \frac{E^{I\,x}}{2}\}\}$$

So for a real number x, you see that
$$\sin[x] = \frac{i\left(e^{-ix} - e^{ix}\right)}{2}$$
and
$$\cos[x] = \frac{e^{ix} + e^{-ix}}{2}$$

To get $\sin[z]$ and $\cos[z]$ for a complex number $z = x + iy$, you just replace x by z in the formulas above to get:
$$\sin[z] = \frac{i\left(e^{-iz} - e^{iz}\right)}{2}$$
and
$$\cos[z] = \frac{e^{iz} + e^{-iz}}{2}$$

Check it out:

In[97]:=
```
N[{Cos[z],(E^(I z) + E^(-I z))/2}/.z->1 + 2 I]
```

Out[97]=
```
{2.03272 - 3.0519 I, 2.03272 - 3.0519 I}
```

In[98]:=
```
N[{Cos[z],(E^(I z) + E^(-I z))/2}/.z->Pi/2]
```

Out[98]=
```
{0, 0}
```

In[99]:=
```
N[{Cos[z],(E^(I z) + E^(-I z))/2}/.z->3 - 2 I]
```

Out[99]=
```
{-3.72455 + 0.511823 I, -3.72455 + 0.511823 I}
```
In[100]:=
```
N[{Sin[z],I(E^(-I z) - E^(I z))/2}/.z-> 1+2 I]
```
Out[100]=
```
{3.16578 + 1.9596 I, 3.16578 + 1.9596 I}
```
In[101]:=
```
N[{Sin[z],I(E^(-I z) - E^(I z))/2}/.z->Pi/2]
```
Out[101]=
```
{1., 1.}
```
In[102]:=
```
N[{Sin[z],I(E^(-I z) - E^(I z))/2}/.z-> 3-2 I]
```
Out[102]=
```
{0.530921 + 3.59056 I, 0.530921 + 3.59056 I}
```

It works every time.

Tutorials

■ T.1) Using expansions to calculate limits

T.1.a.i) Use expansions to calculate
$$\lim_{x \to 0} \frac{f[x]}{g[x]}$$
for $f[x] = 1 - \cos[x]$ and $g[x] = e^{x^2} - 1$.

Answer: Go for it!

In[1]:=
```
Clear[f,g,x]; f[x_] = 1 - Cos[x]; g[x_] = E^(x^2) - 1;
fexpan = Series[f[x],{x,0,4}]
```
Out[1]=
$$\frac{x^2}{2} - \frac{x^4}{24} + O[x]^5$$

In[2]:=
```
gexpan = Series[g[x],{x,0,4}]
```
Out[2]=
$$x^2 + \frac{x^4}{2} + O[x]^5$$

Lesson 4.03 Using Expansions

Divide x^2 into $x^2/2$ to get

$$\lim_{x\to 0} \frac{f[x]}{g[x]} = \lim_{x\to 0} \frac{(x^2/2)}{x^2} = \lim_{x\to 0} \frac{1}{2} = \frac{1}{2}.$$

Check:

In[3]:=
```
Limit[f[x]/g[x],x->0]
```
Out[3]=
$$\frac{1}{2}$$

Mathola.

T.1.a.ii) For a fixed constant b, give a systematic procedure to use to calculate
$$\lim_{x\to b} \frac{f[x]}{g[x]}.$$

Answer: Cut the B.S. and just do it. For $f[x] = e^x \sin[\pi x]^2$ and $g[x] = (1-x)\cos[\pi x/2]$ and $b = 1$:

In[4]:=
```
Clear[f,g,x]; f[x_] = E^x Sin[Pi x]^2;
g[x_] = (1 - x) Cos[Pi x/2]; b = 1;
fexpan = Series[f[x],{x,b,4}]
```
Out[4]=
$$E\, Pi^2\, (-1+x)^2 + E\, Pi^2\, (-1+x)^3 + \left(\frac{E\, Pi^2}{2} - \frac{E\, Pi^4}{3}\right)(-1+x)^4 + O[-1+x]^5$$

In[5]:=
```
gexpan = Series[g[x],{x,b,4}]
```
Out[5]=
$$\frac{Pi\,(-1+x)^2}{2} - \frac{Pi^3\,(-1+x)^4}{48} + O[-1+x]^5$$

Divide $\pi(x-1)^2/2$ into $e\pi^2(x-1)^2$ to get

$$\lim_{x\to 1} \frac{f[x]}{g[x]} = \lim_{x\to 1} \frac{e\,\pi^2\,(x-1)^2}{\pi\,(x-1)^2/2} = \lim_{x\to 1}(2\,e\,\pi) = 2\,e\,\pi.$$

Check:

In[6]:=
```
Limit[f[x]/g[x],x->1]
```
Out[6]=
2 E Pi

Cake.

T.1.a.iii)
> Use expansions to calculate
> $$\lim_{x \to \pi/4} \frac{f[x]}{g[x]}$$
> for $f[x] = \cos[x] - \sin[x]$ and $g[x] = \log[\tan[x]]$.

Answer: Here you go:

In[7]:=
```
Clear[f,g,x]; f[x_] = Cos[x] - Sin[x];
g[x_] = Log[Tan[x]]; b = Pi/4;
fexpan = Series[f[x],{x,b,4}]
```

Out[7]=
$$-(\text{Sqrt}[2]\,(\tfrac{-\text{Pi}}{4} + x)) + \frac{(\tfrac{-\text{Pi}}{4} + x)^3}{3\,\text{Sqrt}[2]} + O[\tfrac{-\text{Pi}}{4} + x]^5$$

In[8]:=
```
gexpan = Series[g[x],{x,b,4}]
```

Out[8]=
$$2\,(\tfrac{-\text{Pi}}{4} + x) + \frac{4\,(\tfrac{-\text{Pi}}{4} + x)^3}{3} + O[\tfrac{-\text{Pi}}{4} + x]^5$$

Divide $2(x - \pi/4)$ into $-\sqrt{2}(x - \pi/4)$ to get

$$\lim_{x \to \pi/4} \frac{f[x]}{g[x]} = \lim_{x \to \pi/4} \frac{-\sqrt{2}(x - \pi/4)}{2(x - \pi/4)} = \lim_{x \to \pi/4} \frac{-\sqrt{2}}{2} = -\frac{1}{\sqrt{2}}.$$

Check:

In[9]:=
```
Limit[f[x]/g[x],x->Pi/4]
```

Out[9]=
$$-(\frac{1}{\text{Sqrt}[2]})$$

Perfecto. You can do this by eyeball faster than *Mathematica* can do this on some machines.

■ T.2) Square roots by Newton's method

T.2.a) Newton's method gives a simple method for extracting square roots on a very cheap calculator.

Here is the idea: Given a number A, you try to calculate $\sqrt{A}$ by going with

$$f[x] = x^2 - A,$$

and using Newton's method to find an approximation of the number x^* that makes $f[x^*] = 0$.

> Say what the update formula is and then say how to punch the appropriate numbers into the cheap calculator.

Answer: Set the function:

In[10]:=
```
Clear[f,a,n,A]
f[x_] = x^2 - A
```
Out[10]=
```
        2
-A + x
```

Take the formula

$$a[n] = a[n-1] - \frac{f[a[n-1]]}{f'[a[n-1]]}$$

from B.2.c) above and clean it up:

In[11]:=
```
Clear[a]; Apart[a[n-1] - f[a[n-1]]/f'[a[n-1]]]
```
Out[11]=
```
    A         a[-1 + n]
─────────── + ─────────
2 a[-1 + n]       2
```

So the clean update formula to punch into the cheap calculator is:

$$a[n] = \left(\frac{1}{2}\right)\left(\frac{A}{a[n-1]} + a[n-1]\right)$$

To illustrate how to do it to approximate $\sqrt{3}$, just punch the following numbers into the calculator, going with a first guess of 1.5:

In[12]:=
```
(1/2) (3/1.5 + 1.5)
```
Out[12]=
```
1.75
```

Now punch in:

In[13]:=
```
(1/2) (3/1.75 + 1.75)
```
Out[13]=
```
1.73214
```

And now punch in:

In[14]:=
```
(1/2) (3/1.73214 + 1.73214)
```

Out[14]=
1.73205

Here is $\sqrt{3}$ to machine accuracy:

In[15]:=
N[Sqrt[3]]

Out[15]=
1.73205

On the calculator, you would have gotten this in three steps. To approximate $\sqrt{10}$, just punch the following numbers into the calculator, going with a first guess of 3.0:

In[16]:=
(1/2) (10/3.0 + 3.0)

Out[16]=
3.16667

The second approximation is:

In[17]:=
(1/2) (10/3.16667 + 3.16667)

Out[17]=
3.16228

The third approximation is:

In[18]:=
(1/2) (10/3.16228 + 3.16228)

Out[18]=
3.16228

Here is $\sqrt{10}$ to machine accuracy:

In[19]:=
N[Sqrt[10]]

Out[19]=
3.16228

On the calculator, you would have gotten this in two steps.

■ T.3) Using the complex exponential to generate trigonometric identities

This problem appears only in the electronic version.

■ T.4) Using expansions to get precise estimates of integrals

T.4.a.i) Here is an attempt to calculate
$$\int_0^{1/2} \sin[x^3]\, dx :$$

Lesson 4.03 Using Expansions

In[20]:=
```
Clear[x]; Integrate[Sin[x^3],{x,0,1/2}]
```

Out[20]=
$$\text{Integrate}[\text{Sin}[x^3], \{x, 0, \tfrac{1}{2}\}]$$

No dice. *Mathematica*'s integrator failed, and so will every other human or machine integrator fail on this. But you can use expansions to estimate the integral $\int_0^{1/2} \sin[x^3]\, dx$ with an error no bigger than 10^{-8}.

Do it.

Answer: Here's the idea:

Use expansions to find a polynomial $p[x]$ that approximates $\sin[x^3]$ within 2×10^{-8} all the way as x runs from 0 to $1/2$. Once you have this $p[x]$, then you'll know that $\int_0^{1/2} p[x]\, dx$ estimates $\int_0^{1/2} \sin[x^3]\, dx$ with an error no bigger than 10^{-8} because:

$$\left| \int_0^{1/2} p[x]\,dx - \int_0^{1/2} \sin[x^3]\,dx \right| = \left| \int_0^{1/2} (p[x] - \sin[x^3])\,dx \right|$$

$$\leq \int_0^{1/2} \left| p[x] - \sin[x^3] \right|\,dx$$

$$\leq \int_0^{1/2} 2 \times 10^{-8}\,dx$$

$$= 2 \times 10^{-8} \left(\frac{1}{2}\right) = 10^{-8}.$$

A reasonable first guess for $p[x]$ is:

In[21]:=
```
Clear[p,x]
p[x_] = Normal[Series[Sin[x],{x,0,3}]]/.x->x^3
```

Out[21]=
$$x^3 - \frac{x^9}{6}$$

Test it at the endpoints:

In[22]:=
```
N[{(p[x] - Sin[x^3])/.x->0, (p[x] - Sin[x^3])/.x->1/2},12]
```

Out[22]=
$$\{0, -2.54218561022\ 10^{-7}\}$$

Close, but no cigar. The error at the right endpoint is bigger than 2×10^{-8}. Respond by going after a better $p[x]$:

In[23]:=
```
Clear[p,x]
p[x_] = Normal[Series[Sin[x],{x,0,5}]]/.x->x^3
```
Out[23]=

$$x^3 - \frac{x^9}{6} + \frac{x^{15}}{120}$$

Test it at the endpoints:

In[24]:=
```
N[{(p[x] - Sin[x^3])/.x->0, (p[x] - Sin[x^3])/.x->1/2},12]
```
Out[24]=

$$\{0,\ 9.45900163751\ 10^{-11}\}$$

Looks good; give it the acid test by plotting the difference between this $p[x]$ and $\sin[x^3]$ all the way from 0 to 1/2:

In[25]:=
```
Clear[difference]
difference[x_] = p[x] - Sin[x^3];
Plot[difference[x],
{x,0,1/2},PlotStyle->
{{Blue,Thickness[0.01]},Red,Red},
PlotRange->All,AxesLabel->{"x","difference x"}];
```

Good. As x advances from 0 to 1/2, this $p[x]$ always runs well within 2×10^{-8} of $\sin[x^3]$.

Now you can announce with confidence and authority that the following number estimates the integral $\int_0^{1/2} \sin[x^3]\, dx$ with an error no bigger than 10^{-8}:

In[26]:=
```
estimate = Integrate[p[x],{x,0,1/2}]
```
Out[26]=

$$\frac{1964033}{125829120}$$

Or, in decimals:

In[27]:=
```
N[estimate,10]
```
Out[27]=
0.01560873191

Only the first 8 decimals are guaranteed to be accurate.

T.4.a.ii) | Why bother with expansions when you can use NIntegrate to estimate
$$\int_0^{1/2} \sin[x^3]\,dx?$$

Answer: Good question. Try it:

In[28]:=
```
NIntegrate[Sin[x^3],{x,0,1/2}]
```

Out[28]=
0.0156087

Compare:

In[29]:=
```
N[estimate,10]
```

Out[29]=
0.01560873191

NIntegrate gave the correct answer to 7 decimals. You can get more accuracy from NIntegrate:

In[30]:=
```
NIntegrate[Sin[x^3],{x,0,1/2},WorkingPrecision->20]
```

Out[30]=
0.015608731903

This is probably even more accurate than the estimate from part T.4.a.i):

In[31]:=
```
N[estimate,10]
```

Out[31]=
0.01560873191

The operative word above is "probably." When you use NIntegrate, the answer you get is probably accurate, but as Stephen Wolfram notes in his book *Mathematica, A System for Doing Mathematics by Computer*, Addison-Wesley, 1991, p.684, "If you have a sufficiently pathological (function to integrate) ... *Mathematica* may simply give you the wrong answer for the integral." The upshot: For practical purposes, NIntegrate can be trusted almost always, but not always. The estimate arrived at in part T.4.a.i) is indisputable.

Give It a Try

Experience with the starred (⋆) problems will be especially beneficial for understanding later lessons.

■ G.1) Approximation of functions by means of their expansions in powers of $(x-b)^\star$

Here is an attempt to get a good approximation for $0 < x < 2$ of
$$f[x] = 2\sin[\pi x] - \cos[\pi x]$$
by means of its expansion in powers of $(x-1)$ through the $(x-1)^6$ term:

In[1]:=
```
b = 1; Clear[f,approx,x];
f[x_] = 2 Sin[Pi x] - Cos[Pi x];
approx[x_] = Normal[Series[f[x],{x,b,6}]]
```

Out[1]=
$$1 - 2\,\text{Pi}\,(-1 + x) - \frac{\text{Pi}^2\,(-1 + x)^2}{2} + \frac{\text{Pi}^3\,(-1 + x)^3}{3} + \frac{\text{Pi}^4\,(-1 + x)^4}{24} - \frac{\text{Pi}^5\,(-1 + x)^5}{60} - \frac{\text{Pi}^6\,(-1 + x)^6}{720}$$

In[2]:=
```
Plot[{f[x],approx[x]},
{x,b - 1,b + 1},PlotStyle->
{{Thickness[0.02],Blue},
{Thickness[0.01],Red}},
AxesLabel->{"x",""},PlotLabel->
"Powers of (x - 1) centered at x = 1",
Epilog->{{Blue,PointSize[0.04],Point[{b,0}]},
Text["expansion point",{b,0},{0,-2}]}];
```

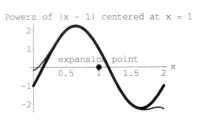

Not great. Here's another attempt using more of the expansion in powers of $(x-1)$:

In[3]:=
```
Clear[approx]
approx[x_] = Normal[Series[f[x],{x,b,10}]];
Plot[{f[x],approx[x]},{x,b - 1,b + 1},
PlotStyle->{{Thickness[0.02],Blue},
{Thickness[0.01],Red}},
AxesLabel->{"x",""},PlotLabel->
"Powers of (x - 1) centered at x = 1",
Epilog->{{Blue,PointSize[0.04],Point[{b,0}]},
Text["expansion point",{b,0},{0,-2}]}];
```

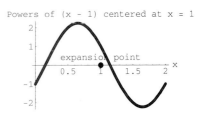

Sharing ink all the way from $x = b-1$ to $x = b+1$. That's a decent approximation.

G.1.a)

Plot
$$f[x] = \sin[\pi x] - 2\cos[2\pi x]$$

and enough of its expansion in powers of $(x-2)$ to see the two plots sharing ink all the way from $x=1$ to $x=3$.

G.1.b) Plot
$$f[x] = e^x$$
and enough of its expansion in powers of $(x-3)$ to see the two plots sharing ink all the way from $x=2$ to $x=4$.

G.1.c) Plot
$$f[x] = x + e^{-0.3x}\cos[6.8\,x]$$
and enough of its expansion in powers of $(x-5)$ to see the two plots sharing ink all the way from $x=4$ to $x=6$.

■ G.2) Calculating some limits★

G.2.a) Use expansions to help calculate
$$\lim_{x \to 0} \frac{\arctan[x^3]}{x\,\sin[x^2]}.$$

G.2.b) Use expansions to help calculate:
$$\lim_{x \to 0} \frac{\sin[a\,x]}{b\,x}, \qquad \lim_{x \to 0} \frac{\sin[a\,x]}{\sin[b\,x]},$$
$$\lim_{x \to 0} \frac{\sin[a^2\,x^2]}{b^2\,x^2}, \qquad \lim_{x \to 0} \frac{\sin[a^2\,x^2]}{\sin[b^2\,x^2]},$$
$$\lim_{x \to 0} \frac{\sin[a\,x]^2}{\sin[b^2\,x^2]}, \qquad \lim_{x \to 0} \frac{\sin[a^3\,x^3]}{\sin[b\,x]^3},$$
and
$$\lim_{x \to 0} \frac{\sin[a^3\,x^3]}{(b\,x)^3}.$$
Discuss the outcomes.

G.2.c.i) Use expansions to help calculate
$$\lim_{x \to 1} \frac{\log[x]}{x - \sqrt{x}}.$$

G.2.c.ii) Use expansions to help calculate
$$\lim_{x \to 2} \frac{x^8 - 256}{x^5 - 32}.$$

G.2.c.iii) Use expansions to help calculate
$$\lim_{x \to 7} \frac{2 - \sqrt{x - 3}}{x^2 - 49}.$$

G.2.c.iv) Use expansions to help calculate
$$\lim_{x \to \pi} \frac{\tan[x/2]^3 - 1}{\cos[x/2]^3}.$$

G.2.d) Put
$$f[x] = \arctan[\arcsin[x]] - \arcsin[\arctan[x]]$$
and put
$$g[x] = \tan[\sin[x]] - \sin[\tan[x]].$$
Use expansions to help calculate
$$\lim_{x \to 0} \frac{f[x]}{g[x]}.$$

■ G.3) Centering the expansion★

G.3.a.i) Which of the following sixth degree polynomials do you predict is the best overall approximation on $[0, 1]$ of $f[x] = e^x$?

```
Clear[x]
approx1[x_] = Normal[Series[E^x,{x,0,6}]]
approx2[x_] = Normal[Series[E^x,{x,1/2,6}]]
approx3[x_] = Normal[Series[E^x,{x,1,6}]]
```

Lesson 4.03 Using Expansions

> Try to explain why the winner had an unfair advantage.

G.3.a.ii) Which of the following do you predict to be the best estimate of $\int_0^1 e^{x^3} dx$? Say why you can predict the right one before you look at the calculations.

```
N[Integrate[approx1[x^3],{x,0,1}],8]
N[Integrate[approx2[x^3],{x,0,1}],8]
N[Integrate[approx3[x^3],{x,0,1}],8]
NIntegrate[E^(x^3),{x,0,1},WorkingPrecision->20]
```

G.3.a.iii) Which of the following do you expect to be the best approximation of a well-behaved function $f[x]$ for $b \le x \le b+1$?

```
Clear[f,b,x]
Normal[Series[f[x],{x, b, 6}]]
Normal[Series[f[x],{x, b + 1/2, 6}]]
Normal[Series[f[x],{x, b + 1, 6}]]
```

Illustrate with some examples and plots.

G.3.b.i) When you go with $b = 20$, is the expansion of e^x in powers of $(x - b)$ through the $(x-b)^8$ term a reasonably good approximation of e^x for $b-1 \le x \le b+1$?

G.3.b.ii) When you go with $b = 1/2$, is the expansion of e^x in powers of $(x - b)$ through the $(x-b)^8$ term a reasonably good approximation of e^x for $b-1 \le x \le b+1$?

G.3.b.iii) Do a few more experiments in the style of parts G.3.b.i) and G.3.b.ii) to help to form an opinion about how the position of the number b on the x-axis influences the quality of the approximation of e^x by the expansion of e^x in powers of $(x - b)$ through the $(x-b)^8$ term on the interval $[b-1, b+1]$.

■ G.4) Root Finding and Newton's method

G.4.a.i) Find two approximate solutions of $\cos[x] = x^2$ by plotting $\cos[x]$ and x^2 on the same axes, and then applying Newton's method to the function
$$f[x] = \cos[x] - x^2$$

at the initial guesses you read off the plot. Confirm your work with the *Mathematica* instruction FindRoot.

Are you sure that there are no more than two solutions? Why?

G.4.a.ii) Use FindRoot to find an approximate solution of
$$e^{-x} = \frac{x^3}{3} - 1.$$
Get your starting point by plotting e^{-x} and $x^3/3 - 1$ on the same axes.

Are you sure that there are no more solutions than the one you found? Why?

G.4.b) Newton's Method can go crazy. When you try to find the root of $x^{1/3} = 0$ by Newton's method, things get a little nasty.

In[4]:=
```
Clear[x]; FindRoot[x^(1/3) == 0,{x,0.2}]
```
FindRoot::cvnwt:
　Newton's method failed to converge to the
　　prescribed accuracy after 15 iterations.

Out[4]=
```
{x -> -0.000390625}
```

Look at this plot of $x^{1/3}$ on $[-10, 10]$ and see if you can guess why this happened.

In[5]:=
```
Plot[Abs[x]^(1/3) Sign[x],{x,-10,10},
  PlotStyle->
    {{Blue,Thickness[0.01]}},
  AxesLabel->{"x","x^(1/3)"}];
```

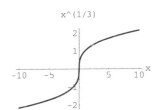

G.4.c.i) Take the update formula
$$a[n] = \left(\frac{1}{3}\right)\left(\frac{A}{a[n-1]^2}\right) + \left(\frac{2}{3}\right)a[n-1]$$
and plug in $A = 8$ and $a[1] = 3$:

In[6]:=
```
Clear[a,n]; A = 8; a[1] = 3;
a[n_] := (1/3)(A/a[n - 1]^2) + (2/3) a[n - 1];
Table[{N[a[n]],N[A^(1/3)]},{n,1,5}]
```

Lesson 4.03 Using Expansions

Out[6]=
{{3., 2.}, {2.2963, 2.}, {2.03659, 2.}, {2.00065, 2.}, {2., 2.}}

Take the same update formula

$$a[n] = \left(\frac{1}{3}\right)\left(\frac{A}{a[n-1]^2}\right) + \left(\frac{2}{3}\right) a[n-1]$$

and plug in $A = 27$ and $a[1] = 4$:

In[7]:=
```
Clear[a,n]; A = 27; a[1] = 4;
a[n_] := (1/3)(A/a[n - 1]^2) + (2/3) a[n - 1];
Table[{N[a[n]],N[A^(1/3)]},{n,1,5}]
```

Out[7]=
{{4., 3.}, {3.22917, 3.}, {3.01588, 3.}, {3.00008, 3.}, {3., 3.}}

Again take the same update formula and plug in $A = 39$ and $a[1] = 4$:

In[8]:=
```
Clear[a,n]; A = 39; a[1] = 4;
a[n_] := (1/3)(A/a[n - 1]^2) + (2/3) a[n - 1];
Table[{N[a[n]],N[A^(1/3)]},{n,1,5}]
```

Out[8]=
{{4., 3.39121}, {3.47917, 3.39121}, {3.39342, 3.39121}, {3.39121, 3.39121}, {3.39121, 3.39121}}

> What does this update do?
>
> Why is this formula handy to use on a cheap calculator that cannot find roots?
>
> Where did the formula come from?

G.4.c.ii) Make up a simple calculator routine to calculate approximate fourth roots on a cheap calculator that cannot find roots.

G.4.c.iii) Design a simple update formula and a calculator routine to use for calculating approximate fifth roots on a cheap calculator that cannot find roots.

G.4.c.iv) Given a positive integer k, what do you think is a simple update formula for calculating an approximate kth root of a number on a cheap calculator that cannot find roots?

■ G.5) Using expansions to get precise estimates of integrals

G.5.a) Here is an attempt to calculate

$$\int_0^{1/5} 10\sqrt{1+\sin[x]^2}\,dx:$$

In[9]:=
```
Clear[x]
Integrate[10 Sqrt[1 + Sin[x]^2],{x,0,1/5}]
```

Out[9]=
$$10\,\text{Integrate}[\text{Sqrt}[1+\sin[x]^2],\,\{x,\,0,\,\tfrac{1}{5}\}]$$

Abort when you get tired of waiting.

> Use expansions to estimate the integral
>
> $$\int_0^{1/5} 10\sqrt{1+\sin[x]^2}\,dx$$
>
> within an error of less than 10^{-10}.
>
> Compare your estimate with the results from NIntegrate.

■ G.6) Expansions and the controversy between spherical mirrors and parabolic mirrors

G.6.a) Here are plots of $f[x] = \sin[x]$, $g[x] = x$, and $h[x] = \tan[x]$ on an interval centered at $x = 0$:

In[10]:=
```
Clear[f,g,h,x]; f[x_] = Sin[x];
g[x_] = Tan[x]; h[x_] = x;
Plot[{f[x],g[x],h[x]},{x,-0.7,0.7},
PlotStyle->{{Thickness[0.01],SkyBlue},
{Thickness[0.01],Brown},
{Thickness[0.005],Red}},
AxesLabel->{"x",""}];
```

> Use expansions to help explain why the plots turned out this way, thereby explaining why scientists often do not distinguish among $f[x] = \sin[x]$, $g[x] = \tan[x]$, and $h[x] = x$ when x is close to 0.

G.6.b) If you've been around Calculus&Mathematica from the beginning, then you probably know that parabolic mirrors shaped like this concentrate vertical light rays at a single focal point:

In[11]:=
    Clear[x]
    Plot[-4 + (x^2)/3,{x,-2,2},
    PlotStyle->{{Blue,Thickness[0.01]}},
    AxesLabel->{"x",""},AspectRatio->Automatic];

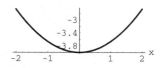

Using tan[x] and x interchangeably for x's close to 0, the Sears, Zemansky, and Young freshman physics text, *University Physics* (Addison-Wesley, 1987, p.871), comes to the conclusion that some spherical mirrors also concentrate vertical light rays at a single focal point. They imply that a big spherical mirror like this will not concentrate vertical light rays at a single focal point:

In[12]:=
    Plot[-Sqrt[4 - x^2],{x,-2,2},
    PlotStyle->{{Blue,Thickness[0.01]}},
    AxesLabel->{"x",""},
    AspectRatio->Automatic];

But they imply that this part of the spherical mirror above will concentrate vertical light rays at a single focal point:

In[13]:=
    Plot[-Sqrt[4 - x^2],{x,-0.4,0.4},
    PlotStyle->{{Blue,Thickness[0.01]}},
    PlotRange->{{-2,2},{-2,0}},
    AxesLabel->{"x",""},
    AspectRatio->Automatic];

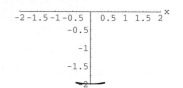

Here are the mathematical facts of the matter:

→ No part of a spherical mirror ever concentrates vertical light rays at a single focal point.

→ All parabolic mirrors like the one above do concentrate vertical light rays at a single focal point. Look at the following:

In[14]:=
    Clear[expan2,x]
    expan2[x_] = Normal[Series[-Sqrt[4 - x^2],{x,0,2}]]

Out[14]=
    $-2 + \dfrac{x^2}{4}$

This is the formula of a parabola. Now look at the following plots of the spherical mirror above and a closely related parabolic mirror:

```
In[15]:=
    Plot[{-Sqrt[4 - x^2],expan2[x]},{x,-2,2},
    PlotStyle->{{Blue,Thickness[0.02]},
    {Red,Thickness[0.01]}},
    PlotRange->{{-2,2},{-2,0}},
    AxesLabel->{"x",""},AspectRatio->Automatic];
```

> Eyeball this and then plot the part of the spherical mirror that you believe will do a pretty good job of concentrating vertical light rays at a single point.
>
> If you know how to bounce light off mirrors, run some experiments.
>
> Does this begin to resolve the controversy between parabolic and spherical mirrors?

■ G.7) Getting the expansion of tan[x] by division

G.7.a) Trying to come up with the formula for the general term of the expansion of tan[x] in powers of x is not easy.

> Compromise by coming up with the expansion of tan[x] in powers of x through the x^9 term, by dividing enough of the expansion of cos[x] in powers of x into enough of the expansion of sin[x] in powers of x.

■ G.8) Generating identities for sin[m x]

This problem appears only in the electronic version.

■ G.9) Behavior close to 0★

G.9.a)
> Plot $\sin[x]^2$, x^2, and $\sin[x^2]$ on the same axes for $-1 \le x \le 1$.
>
> What is it about the expansions of each of these functions that might explain the plots?

G.9.b)
> Plot $\sin[x]^2$ and $x^2 (\cos[x])^{2/3}$ on the same axes for $-1 \le x \le 1$.
>
> What is it about the expansions of each of these functions that might explain the plots?

Lesson 4.03 Using Expansions

G.9.c) Plot

$$f[x] = \sin[x] \quad \text{and} \quad g[x] = \frac{x\,(60 - 7\,x^2)}{60 + 3\,x^2}$$

on the same axes for $-3 \le x \le 3$. What is it about the expansions of each of these functions that might explain the plots?

G.9.d) What part of the expansion of a function $f[x]$ in powers of x best reflects the behavior of the function for x's close to 0?

G.9.e) Set a constant c so that the plots of $\cos[x]$ and e^{-cx^2} share lots of ink on small intervals centered at 0.

Back up your answer with a plot.

■ G.10) Behavior away from 0*

G.10.a) Look at:

In[16]:=
```
Clear[expan3,expan5,x]
expan3[x_] = Normal[Series[Sin[x],{x,0,3}]]
```

Out[16]=
$$x - \frac{x^3}{6}$$

In[17]:=
```
expan5[x_] = Normal[Series[Sin[x],{x,0,5}]]
```

Out[17]=
$$x - \frac{x^3}{6} + \frac{x^5}{120}$$

In[18]:=
```
Plot[{Sin[x],expan3[x],expan5[x]},{x,-4,4},
PlotStyle->{{Thickness[0.02]},
{Thickness[0.015],Blue},{Thickness[0.01],Red}},
PlotRange->{-2.5,2.5},AxesLabel->{"x",""}];
```

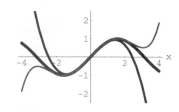

Identify each of the plots.

Look at the sign of the highest degree term in each of the partial expansions and discuss how it influences the plot of each partial expansion on the far right.

How about on the far left?

G.10.b) Look at:

In[19]:=
```
Clear[expan8,expan10,x]
expan8[x_] = Normal[Series[E^(-x^2),{x,0,8}]]
```

Out[19]=

$$1 - x^2 + \frac{x^4}{2} - \frac{x^6}{6} + \frac{x^8}{24}$$

In[20]:=
```
expan10[x_] = Normal[Series[E^(-x^2),{x,0,10}]]
```

Out[20]=

$$1 - x^2 + \frac{x^4}{2} - \frac{x^6}{6} + \frac{x^8}{24} - \frac{x^{10}}{120}$$

In[21]:=
```
Plot[{E^(-x^2),expan8[x],expan10[x]},{x,-2,2},
PlotStyle->{{Thickness[0.02]},
{Thickness[0.015],Blue},{Thickness[0.01],Red}},
PlotRange->{-2.5,2.5},AxesLabel->{"x",""}];
```

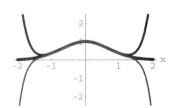

Identify each of the plots.

Look at the sign of the highest degree term in each of the partial expansions and discuss how it influences the plot of each partial expansion on the far right.

How about on the far left?

G.10.c) Look at:

In[22]:=
```
Clear[x]
expansion10 = Normal[Series[ArcTan[x^2],{x,0,10}]]
```

Out[22]=

$$x^2 - \frac{x^6}{3} + \frac{x^{10}}{5}$$

In[23]:=
```
expansion14 = Normal[Series[ArcTan[x^2],{x,0,14}]]
```

Out[23]=
$$x^2 - \frac{x^6}{3} + \frac{x^{10}}{5} - \frac{x^{14}}{7}$$

In[24]:=
```
Plot[{ArcTan[x^2],expansion10,expansion14},
 {x,-1.3,1.3},PlotRange->{0,2},
 PlotStyle->{{Thickness[0.005]},
 {Thickness[0.01],Blue},{Thickness[0.01],Red}}];
```

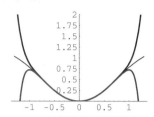

> Identify each of the plots.
>
> Look at the sign of the highest degree term in each of the partial expansions and explain how it influences the plot of each partial expansion on the far right.
>
> How about on the far left?

G.10.d) The only information you have about a certain function $f[x]$ is:

$$-1 \leq f[x] \leq 1$$

for all the x's between $-\infty$ and ∞.

> Is it possible for a plot of a partial expansion of $f[x]$ to share ink with the plot of $f[x]$ all the way from $-\infty$ to $+\infty$? Why?

■ G.11) Error at the endpoints

This problem appears only in the electronic version.

LESSON 4.04

Taylor's Formula

Basics

■ B.1) Taylor's formula

B.1.a) Look at these:

In[1]:=
   `Clear[f,b,x]; Normal[Series[f[x],{x,b,3}]]`

Out[1]=

$$f[b] + (-b + x)\, f'[b] + \frac{(-b + x)^2\, f''[b]}{2} + \frac{(-b + x)^3\, f^{(3)}[b]}{6}$$

In[2]:=
   `Normal[Series[f[x],{x,b,8}]]`

Out[2]=

$$f[b] + (-b + x)\, f'[b] + \frac{(-b + x)^2\, f''[b]}{2} + \frac{(-b + x)^3\, f^{(3)}[b]}{6} +$$

$$\frac{(-b + x)^4\, f^{(4)}[b]}{24} + \frac{(-b + x)^5\, f^{(5)}[b]}{120} + \frac{(-b + x)^6\, f^{(6)}[b]}{720} +$$

$$\frac{(-b + x)^7\, f^{(7)}[b]}{5040} + \frac{(-b + x)^8\, f^{(8)}[b]}{40320}$$

> What's the message?

Answer: Look at some more:

In[3]:=
```
Clear[f,b,x]; Normal[Series[f[x],{x,b,2}]]
```
Out[3]=

$$f[b] + (-b + x) \, f'[b] + \frac{(-b + x)^2 \, f''[b]}{2}$$

This has order of contact 2 with $f[x]$ at $x = b$.

In[4]:=
```
Normal[Series[f[x],{x,b,6}]]
```
Out[4]=

$$f[b] + (-b + x) \, f'[b] + \frac{(-b + x)^2 \, f''[b]}{2} + \frac{(-b + x)^3 \, f^{(3)}[b]}{6} +$$

$$\frac{(-b + x)^4 \, f^{(4)}[b]}{24} + \frac{(-b + x)^5 \, f^{(5)}[b]}{120} + \frac{(-b + x)^6 \, f^{(6)}[b]}{720}$$

This has order of contact 6 with $f[x]$ at $x = b$. The denominators are factorials. The message is that the expansion of $f[x]$ in powers of $(x - b)$ is

$$f[b] + f'[b](x-b) + \frac{f''[b]}{2!}(x-b)^2 + \frac{f^{(3)}[b]}{3!}(x-b)^3 + \cdots + \frac{f^{(k)}[b]}{k!}(x-b)^k + \cdots$$

Most everyone calls this Taylor's formula.

B.1.b) Check out Taylor's formula for the coefficients of the expansion of $f[x] = e^x$ in powers of $(x - 1)$.

Answer: The expansion of e^x in powers of $(x - 1)$ starts out with:

In[5]:=
```
n = 6; b = 1;
Clear[x,f]; f[x] = E^x;
Normal[Series[f[x],{x,b,n}]]
```
Out[5]=

$$E + E \, (-1 + x) + \frac{E \, (-1 + x)^2}{2} + \frac{E \, (-1 + x)^3}{6} +$$

$$\frac{E \, (-1 + x)^4}{24} + \frac{E \, (-1 + x)^5}{120} + \frac{E \, (-1 + x)^6}{720}$$

Taylor's formula for the expansion of $f[x]$ in powers of $(x - 1)$ is

$$f[1] + f'[1](x-1) + \frac{f''[1]}{2!}(x-1)^2 + \frac{f^{(3)}[1]}{3!}(x-1)^3 + \cdots + \frac{f^{(k)}[1]}{k!}(x-1)^k + \cdots$$

So Taylor's formula gives you:

In[6]:=
```
Clear[a,k];
a[k_] := (D[f[x],{x,k}]/.x->b)/k!
Sum[a[k] (x - b)^k,{k,0,n}]
```

Out[6]=

$$E + E\,(-1 + x) + \frac{E\,(-1 + x)^2}{2} + \frac{E\,(-1 + x)^3}{6} + \frac{E\,(-1 + x)^4}{24} + \frac{E\,(-1 + x)^5}{120} + \frac{E\,(-1 + x)^6}{720}$$

Looks good. Do it again:

In[7]:=
```
n = 12; Normal[Series[f[x],{x,b,n}]]
```

Out[7]=

$$E + E\,(-1 + x) + \frac{E\,(-1 + x)^2}{2} + \frac{E\,(-1 + x)^3}{6} + \frac{E\,(-1 + x)^4}{24} + \frac{E\,(-1 + x)^5}{120} +$$
$$\frac{E\,(-1 + x)^6}{720} + \frac{E\,(-1 + x)^7}{5040} + \frac{E\,(-1 + x)^8}{40320} + \frac{E\,(-1 + x)^9}{362880} + \frac{E\,(-1 + x)^{10}}{3628800} +$$
$$\frac{E\,(-1 + x)^{11}}{39916800} + \frac{E\,(-1 + x)^{12}}{479001600}$$

In[8]:=
```
Sum[a[k] (x - b)^k,{k,0,n}]
```

Out[8]=

$$E + E\,(-1 + x) + \frac{E\,(-1 + x)^2}{2} + \frac{E\,(-1 + x)^3}{6} + \frac{E\,(-1 + x)^4}{24} + \frac{E\,(-1 + x)^5}{120} +$$
$$\frac{E\,(-1 + x)^6}{720} + \frac{E\,(-1 + x)^7}{5040} + \frac{E\,(-1 + x)^8}{40320} + \frac{E\,(-1 + x)^9}{362880} + \frac{E\,(-1 + x)^{10}}{3628800} +$$
$$\frac{E\,(-1 + x)^{11}}{39916800} + \frac{E\,(-1 + x)^{12}}{479001600}$$

Play with these by going back to the beginning and changing $f[x]$, b, and n.

B.1.c.i) | What use is Taylor's formula?

Answer: Many new practitioners of calculus want to jump on this formula and use it to slam out all expansions. But those who have been around for a while know this is not a good idea even if you are as fast as *Mathematica*. As a matter

Lesson 4.04 Taylor's Formula

of fact, this formula is usually the least efficient way to obtain an expansion of a given function. Just think of the misery involved in calculating many derivatives and plugging in. Just to use Taylor's formula to calculate the first 5 terms of the expansion of $f[x] = e^x/(1-x)$ in powers of x, you've got to calculate this bird's nest of derivatives:

In[9]:=
```
Clear[f,x]; f[x_] = (E^x)/(1 - x);
Table[D[f[x],{x,k}]/k!,{k,0,5}]
```

Out[9]=

$$\left\{ \frac{e^x}{1-x}, \frac{e^x}{(1-x)^2} + \frac{e^x}{1-x}, \frac{\frac{2e^x}{(1-x)^3} + \frac{2e^x}{(1-x)^2} + \frac{e^x}{1-x}}{2}, \right.$$

$$\frac{\frac{6e^x}{(1-x)^4} + \frac{6e^x}{(1-x)^3} + \frac{3e^x}{(1-x)^2} + \frac{e^x}{1-x}}{6},$$

$$\frac{\frac{24e^x}{(1-x)^5} + \frac{24e^x}{(1-x)^4} + \frac{12e^x}{(1-x)^3} + \frac{4e^x}{(1-x)^2} + \frac{e^x}{1-x}}{24},$$

$$\left. \frac{\frac{120e^x}{(1-x)^6} + \frac{120e^x}{(1-x)^5} + \frac{60e^x}{(1-x)^4} + \frac{20e^x}{(1-x)^3} + \frac{5e^x}{(1-x)^2} + \frac{e^x}{1-x}}{120} \right\}$$

To get the expansion of $f[x] = e^x/(1-x)$ in powers of x through the x^5 term, the experienced practitioner of calculus takes:

In[10]:=
```
first = Series[E^x,{x,0,5}]
```

Out[10]=

$$1 + x + \frac{x^2}{2} + \frac{x^3}{6} + \frac{x^4}{24} + \frac{x^5}{120} + O[x]^6$$

In[11]:=
```
second = Series[1/(1 - x),{x,0,5}]
```

Out[11]=

Out[20]=
$$1 + x + x^2 + x^3 + x^4 + x^5 + O[x]^6$$

And multiplies them together:

In[12]:=
```
first second
```

Out[12]=
$$1 + 2x + \frac{5x^2}{2} + \frac{8x^3}{3} + \frac{65x^4}{24} + \frac{163x^5}{60} + O[x]^6$$

Although Taylor's formula is not very useful for down-to-earth computations, it does have quite a bit of theoretical value. And since theory feeds new calculations and measurements, it has some practical value as well. You'll see it again in this lesson.

B.1.c.ii) Does *Mathematica* use Taylor's formula to come up with expansions?

Answer: Only as a last resort. Like most experienced scientists, *Mathematica* tries to get its expansions the way you have been doing it: Memorize basic expansions like the expansions of $1/(1-x)$, $\sin[x]$, $\cos[x]$, and e^x in powers of x and then do little adjustments to get the expansions you want.

B.1.d) How does Taylor's formula guarantee that it is O.K. to get expansions by devices other than Taylor's formula?

Answer: This is a good example of the use of Taylor's formula as a theoretical tool. Taylor's formula tells you that there is a definite formula for expansions.

In[13]:=
`Clear[f]; Series[f[x],{x,0,7}]`

Out[13]=
$$f[0] + f'[0]\,x + \frac{f''[0]\,x^2}{2} + \frac{f^{(3)}[0]\,x^3}{6} + \frac{f^{(4)}[0]\,x^4}{24} + \frac{f^{(5)}[0]\,x^5}{120} + \frac{f^{(6)}[0]\,x^6}{720} + \frac{f^{(7)}[0]\,x^7}{5040} + O[x]^8$$

The fact that there is such a formula means that there is only one possibility for an expansion in powers of x. As a result, you always get the same expansion for a given function no matter how you go about it. So get it any way you can!

■ B.2) Four approximations based on Taylor's formula

Calculus&*Mathematica* is very pleased to acknowledge that the presentation here is greatly influenced by the book *Numerical Methods and Software* by David Kahaner, Cleve Moler, and Stephen Nash, Prentice-Hall, 1989. This book is a winner.

Euler approximation: Here is what some folks call the Euler approximation of a function $f[x]$ at a point $x = b$:

Lesson 4.04 Taylor's Formula

$In[14]:=$
```
Clear[feuler,f,x,b]
feuler[x_,b_] = f[b] + (x - b) f'[b]
```
$Out[14]=$
f[b] + (-b + x) f'[b]

Check the order of contact at $x = b$:

$In[15]:=$
```
{Series[feuler[x,b],{x,b,3}],Series[f[x],{x,b,3}]}
```
$Out[15]=$
{f[b] + f'[b] (-b + x) + O[-b + x]^4,

f[b] + f'[b] (-b + x) + $\dfrac{f''[b] (-b + x)^2}{2}$ + $\dfrac{f^{(3)}[b] (-b + x)^3}{6}$ + O[-b + x]^4 }

Good. The functions feuler$[x, b]$ and $f[x]$ have order of contact 1 at $x = b$. Take a look at a plot in the specific case of

$$f[x] = x + \frac{e^x \sin[4\pi x]}{10} \quad \text{and} \quad b = 0:$$

$In[16]:=$
```
Clear[f]
f[x_] = x + (E^x) Sin[4 Pi x]/10;
Plot[{f[x],feuler[x,0]},{x,0,1/2},
PlotStyle->{{Blue,Thickness[0.02]},
{Red,Thickness[0.01]}},PlotRange->{0,0.8},
AxesLabel->{"x",""},PlotLabel->
"Euler Approximation"];
```

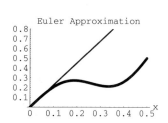

Trapezoidal approximation: Here is what some folks call the trapezoidal approximation of a function $f[x]$ at a point $x = b$:

$In[17]:=$
```
Clear[ftrap,f,x,b]
ftrap[x_,b_] = f[b] + (x - b) (f'[b] + f'[x])/2
```
$Out[17]=$
f[b] + $\dfrac{(-b + x) (f'[b] + f'[x])}{2}$

Check the order of contact at $x = b$:

$In[18]:=$
```
{Series[ftrap[x,b],{x,b,3}],Series[f[x],{x,b,3}]}
```
$Out[18]=$
{f[b] + f'[b] (-b + x) + $\dfrac{f''[b] (-b + x)^2}{2}$ + $\dfrac{f^{(3)}[b] (-b + x)^3}{4}$ + O[-b + x]^4,

f[b] + f'[b] (-b + x) + $\dfrac{f''[b] (-b + x)^2}{2}$ + $\dfrac{f^{(3)}[b] (-b + x)^3}{6}$ + O[-b + x]^4 }

Even better. ftrap[x, b] and f[x] have order of contact 2 at x = b. This is moderately interesting because ftrap[x, b] makes use of only first order (= first derivative) information. So you get second derivative confidence with only first derivative information. Take a look at a plot in the specific case of

$$f[x] = x + \frac{e^x \sin[4\pi x]}{10} \quad \text{and} \quad b = 0:$$

In[19]:=
```
Clear[f]
f[x_] = x + (E^x) Sin[4 Pi x]/10;
Plot[{f[x],ftrap[x,0]},{x,0,1/2},
PlotStyle->{{Blue,Thickness[0.02]},
{Red,Thickness[0.01]}},PlotRange->{0,0.8},
AxesLabel->{"x",""},PlotLabel->
"Trapezoidal Approximation"];
```

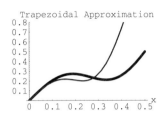

Midpoint approximation: Here is what some folks call the midpoint approximation of a function $f[x]$ at a point $x = b$:

In[20]:=
```
Clear[fmidpt,f,x,b]
fmidpt[x_,b_] = f[b] + (x - b) f'[(b + x)/2]
```

Out[20]=

$$f[b] + (-b + x)\ f'[\frac{b + x}{2}]$$

Check the order of contact at $x = b$:

In[21]:=
```
{Series[fmidpt[x,b],{x,b,3}],
Series[f[x],{x,b,3}]}
```

Out[21]=

$$\{f[b] + f'[b]\ (-b + x) + \frac{f''[b]\ (-b + x)^2}{2} + \frac{f^{(3)}[b]\ (-b + x)^3}{8} + O[-b + x]^4,$$

$$f[b] + f'[b]\ (-b + x) + \frac{f''[b]\ (-b + x)^2}{2} + \frac{f^{(3)}[b]\ (-b + x)^3}{6} + O[-b + x]^4\}$$

Nice. The function fmidpt[x, b] and f[x] have order of contact 2 at x = b. Again, this is moderately interesting because fmidpt[x, b] makes use of only first order (= first derivative) information, so you get second derivative confidence with only first derivative information. Take a look at a plot in the specific case of

$$f[x] = x + \frac{e^x \sin[4\pi x]}{10} \quad \text{and} \quad b = 0:$$

Lesson 4.04 Taylor's Formula

In[22]:=
```
Clear[f]
f[x_] = x + (E^x) Sin[4 Pi x]/10;
Plot[{f[x],fmidpt[x,0]},{x,0,1/2},
PlotStyle->{{Blue,Thickness[0.02]},
{Red,Thickness[0.01]}},PlotRange->{0,0.8},
AxesLabel->{"x",""},PlotLabel->
"Midpoint Approximation"];
```

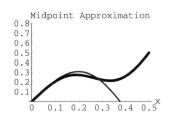

Runge-Kutta approximation: Here is what some folks call a Runge-Kutta approximation of a function $f[x]$ at a point $x = b$:

In[23]:=
```
Clear[frunge,f,x,b]
frunge[x_,b_] = (2 fmidpt[x,b] + ftrap[x,b])/3
```

Out[23]=

$$\frac{f[b] + \frac{(-b + x)\,(f'[b] + f'[x])}{2} + 2\,(f[b] + (-b + x)\,f'[\frac{b + x}{2}])}{3}$$

Check the order of contact at $x = b$:

In[24]:=
```
{Series[frunge[x,b],{x,b,5}],Series[f[x],{x,b,5}]}
```

Out[24]=

$$\{f[b] + f'[b]\,(-b + x) + \frac{f''[b]\,(-b + x)^2}{2} + \frac{f^{(3)}[b]\,(-b + x)^3}{6} +$$

$$\frac{f^{(4)}[b]\,(-b + x)^4}{24} + \frac{5\,f^{(5)}[b]\,(-b + x)^5}{576} + O[-b + x]^6\,,$$

$$f[b] + f'[b]\,(-b + x) + \frac{f''[b]\,(-b + x)^2}{2} + \frac{f^{(3)}[b]\,(-b + x)^3}{6} +$$

$$\frac{f^{(4)}[b]\,(-b + x)^4}{24} + \frac{f^{(5)}[b]\,(-b + x)^5}{120} + O[-b + x]^6\,\}$$

Good deal. The function frunge$[x,b]$ and $f[x]$ have order of contact 4 (YES, 4). This is very interesting because frunge$[x,b]$ makes use of only first order (= first derivative) information. So you get fourth derivative confidence with only first derivative information.

When you realize that 5/576 is close to 1/120, then you realize that frunge$[x,b]$ and $f[x]$ come very close to having order of contact 5 at $x = b$.

Take a look at a plot in the specific case of

$$f[x] = x + \frac{e^x \sin[4\pi x]}{10} \quad \text{and} \quad b = 0:$$

In[25]:=
```
Clear[f]
f[x_] = x + (E^x) Sin[4 Pi x]/10;
Plot[{f[x],frunge[x,0]},{x,0,1/2},
PlotStyle->{{Blue,Thickness[0.02]},
{Red,Thickness[0.01]}},PlotRange->{0,0.8},
AxesLabel->{"x",""},PlotLabel->
"Runge-Kutta Approximation"];
```

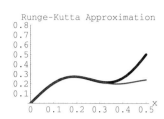

Nice approximation resulting from order of contact 4 at $x = 0$.

B.2.a.i) The Runge-Kutta approximation of $f[x]$ near $x = b$ is
$$\text{frunge}[x, b] = \frac{2\,\text{fmidpt}[x, b] + \text{ftrap}[x, b]}{3}.$$

> What's the idea behind the Runge-Kutta approximation?

Answer: Remember:

In[26]:=
```
Clear[fmidpt,f,x,b]
fmidpt[x_,b_] = f[b] + (x - b) f'[(b + x)/2]
```
Out[26]=
$$f[b] + (-b + x)\, f'[\frac{b + x}{2}]$$

In[27]:=
```
Clear[ftrap]
ftrap[x_,b_] = f[b] + (x - b) (f'[b] + f'[x])/2
```
Out[27]=
$$f[b] + \frac{(-b + x)\,(f'[b] + f'[x])}{2}$$

Take a number t and look at the expansion of the combination
$$t\,\text{fmidpt}[x, b] + (1 - t)\,\text{ftrap}[x, b]$$
in powers of $(x - b)$:

In[28]:=
```
Clear[t]
Series[t fmidpt[x,b] + (1 - t) ftrap[x,b],{x,b,4}]
```
Out[28]=
$$((1 - t)\,f[b] + t\,f[b]) + ((1 - t)\,f'[b] + t\,f'[b])\,(-b + x) +$$
$$\left(\frac{(1 - t)\,f''[b]}{2} + \frac{t\,f''[b]}{2}\right)(-b + x)^2 +$$
$$\left(\frac{(1 - t)\,f^{(3)}[b]}{4} + \frac{t\,f^{(3)}[b]}{8}\right)(-b + x)^3 +$$

Lesson 4.04 Taylor's Formula

$$\left(\frac{(1-t)\,f^{(4)}[b]}{12} + \frac{t\,f^{(4)}[b]}{48}\right)(-b+x)^4 + O[-b+x]^5$$

Look at the first three terms:

In[29]:=
```
first = Expand[(1 - t) f[b] + t f[b]]
```

Out[29]=
```
f[b]
```

In[30]:=
```
second = Factor[((1 - t) f'[b] + t f'[b]) (x - b)]
```

Out[30]=
```
(-b + x) f'[b]
```

In[31]:=
```
third = Factor[((1 - t) f''[b]/2 + t f''[b]/2) (x - b)^2]
```

Out[31]=

$$\frac{(-b + x)^2\, f''[b]}{2}$$

This tells you that no matter what t you go with, you are assured that

$$t\,\text{fmidpt}[x, b] + (1 - t)\,\text{ftrap}[x, b]$$

has order of contact at least 2 with $f[x]$ at $x = b$. Look again:

In[32]:=
```
Clear[t]
Series[t fmidpt[x,b] + (1 - t) ftrap[x,b],{x,b,4}]
```

Out[32]=

$$((1 - t)\,f[b] + t\,f[b]) + ((1 - t)\,f'[b] + t\,f'[b])(-b + x) +$$

$$\left(\frac{(1-t)\,f''[b]}{2} + \frac{t\,f''[b]}{2}\right)(-b + x)^2 +$$

$$\left(\frac{(1-t)\,f^{(3)}[b]}{4} + \frac{t\,f^{(3)}[b]}{8}\right)(-b + x)^3 +$$

$$\left(\frac{(1-t)\,f^{(4)}[b]}{12} + \frac{t\,f^{(4)}[b]}{48}\right)(-b + x)^4 + O[-b + x]^5$$

This time look at the fourth term:

In[33]:=
```
fourth = Factor[((1 - t) f'''[b]/4 + t f'''[b]/8)
    (x - b)^3]
```

Out[33]=

$$\frac{(2 - t)(-b + x)^3\, f^{(3)}[b]}{8}$$

To pick up order of contact 3 at $x = b$, set t so that this is the same as

$$\frac{f^3[b]}{3!}(x-b)^3,$$

look at:

In[34]:=
```
Solve[fourth == (f'''[b]/3!) (x - b)^3,t]
```
Out[34]=
```
{{t -> 2/3}}
```

This tells you that taking $t = 2/3$ in

$$t\,\text{fmidpt}[x,b] + (1-t)\,\text{ftrap}[x,b]$$

guarantees order of contact 3 with $f[x]$ at $x = b$. And when you go with $t = 2/3$, you get

$$\text{frunge}[x,b] = \frac{2\,\text{fmidpt}[x,b] + \text{ftrap}[x,b]}{3}.$$

This is the idea behind the Runge-Kutta approximation.

B.2.a.ii) Hold your horses! The Runge-Kutta approximation

$$\text{frunge}[x,b] = \frac{2\,\text{fmidpt}[x,b] + \text{ftrap}[x,b]}{3}$$

was designed to have order of contact 3 with $f[x]$ at $x = b$. As worked out above, frunge$[x,b]$ actually has order of contact 4 with $f[x]$ at $x = b$.

> Is this some sort of calculational accident?

Answer: Get ready for this. It hardly ever happens. The answer is:

Yes!

It's a fringe benefit.

B.2.b)
> How can you use these approximations to estimate $\int_a^c f'[x]\,dx$?

Answer: Really you shouldn't. You're a lot better off using Integrate, NIntegrate or NDSolve, all of which use professionally written methods more advanced than these approximations. Nevertheless, using these methods gives the flavor of the more advanced methods, so it's not a bad idea to see how they can be used. Here are your toys; activate them.

Euler approximation:

In[35]:=
```
Clear[feuler,f,x,b,Derivative]
feuler[x_,b_] = f[b] + (x - b) f'[b]
```

Out[35]=
```
f[b] + (-b + x) f'[b]
```

Trapezoidal approximation:

In[36]:=
```
Clear[ftrap,f,x,b,Derivative]
ftrap[x_,b_] = f[b] + (x - b) (f'[b] + f'[x])/2
```

Out[36]=

$$f[b] + \frac{(-b + x)\,(f'[b] + f'[x])}{2}$$

Midpoint approximation:

In[37]:=
```
Clear[fmidpt,f,x,b,Derivative]
fmidpt[x_,b_] = f[b] + (x - b) f'[(b + x)/2]
```

Out[37]=

$$f[b] + (-b + x)\,f'[\tfrac{b + x}{2}]$$

Runge-Kutta approximation:

In[38]:=
```
Clear[frunge,f,x,b,Derivative]
frunge[x_,b_] = (2 fmidpt[x,b] + ftrap[x,b])/3
```

Out[38]=

$$\frac{f[b] + \frac{(-b + x)\,(f'[b] + f'[x])}{2} + 2\,(f[b] + (-b + x)\,f'[\tfrac{b + x}{2}])}{3}$$

Given a formula for $f'[x]$, you can use these approximations to estimate the integral $\int_a^c f'[x]\,dx$.

Here's the key observation: Set

$$\text{jump} = \frac{c - a}{n}$$

and note that the exact value of the integral is measured by

$$\text{Sum}[f[b + \text{jump}] - f[b], \{b, a, c - \text{jump}, \text{jump}\}]$$
$$= \text{Sum}[\int_b^{b+\text{jump}} f'[x]\,dx, \{b, a, c - \text{jump}, \text{jump}\}] = \int_a^c f'[x]\,dx.$$

Basics (B.2)

The idea here is to estimate

$$\int_a^c f'[x]\,dx = \text{Sum}[f[b+\text{jump}] - f[b], \{b, a, c-\text{jump}, \text{jump}\}]$$

by calculating

$$\text{Sum}[\text{fapprox}[b+\text{jump}, b] - \text{fapprox}[b, b]], \{b, a, c-\text{jump}, \text{jump}\}]$$

where fapprox$[x, b]$ is an approximation of $f[x]$ that has good contact with $f[x]$ at $x = b$, like the approximations above.

Here's how this idea works to estimate $\int_0^2 f'[x]\,dx$ in the case that $f'[x] = e^{\sin[x]}$. *Mathematica* cannot come up with the exact value of $\int_0^2 f'[x]\,dx$, so go for *Mathematica*'s estimate of $\int_0^2 f'[x]\,dx$:

In[39]:=
```
Clear[f,Derivative,x]; f'[x_] = E^(Sin[x]);
a = 0.0; c = 2.0;
MathematicaEstimate = NIntegrate[f'[x],{x,a,c}]
```

Out[39]=
4.23653

Now set up the calculation

$$\text{Sum}[\text{fapprox}[b+\text{jump}, b] - \text{fapprox}[b, b]], \{b, a, c-\text{jump}, \text{jump}\}]$$

for each of the approximations worked with above:

In[40]:=
```
Clear[b,jump,EulerEstimate,TrapezoidalEstimate,MidpointEstimate,SimpsonEstimate]
EulerEstimate[jump_] := N[Sum[feuler[b + jump,b] - feuler[b,b],{b,a,c - jump,jump}]];
TrapezoidalEstimate[jump_] := N[Sum[
   ftrap[b + jump,b] - ftrap[b,b],{b,a,c - jump,jump}]];
MidpointEstimate[jump_] := N[Sum[
   fmidpt[b + jump,b] - fmidpt[b,b],{b,a,c - jump,jump}]];
RungeKuttaEstimate[jump_] := N[Expand[Sum[
   frunge[b + jump,b] - frunge[b,b],{b,a,c - jump,jump}]]];
jump = (c - a)/4;
ColumnForm[{jump "= jump gives you these estimates:","",
EulerEstimate[jump] "= Euler",
TrapezoidalEstimate[jump] "= Trapezoidal",
MidpointEstimate[jump] "= Midpoint",
RungeKuttaEstimate[jump] "= Runge-Kutta",
MathematicaEstimate "=  MathematicaEstimate"}]
```

Out[40]=
```
0.5 = jump gives you these estimates:

3.8232 = Euler
4.19385 = Trapezoidal
4.258 = Midpoint
4.23661 = Runge-Kutta
4.23653 =   MathematicaEstimate
```

Just as you probably expected, the higher order contact approximations gave the better results. In fact, the Runge-Kutta approximation did a darn good job even

Lesson 4.04 Taylor's Formula

with a big jump size. See what happens when you reduce the jump size from $(c-a)/4$ to $(c-a)/10$:

In[41]:=
```
jump = (c - a)/10;
ColumnForm[{jump "= jump gives you these estimates:",
  "",EulerEstimate[jump] "= Euler",
  TrapezoidalEstimate[jump] "= Trapezoidal",
  MidpointEstimate[jump] "= Midpoint",
  RungeKuttaEstimate[jump] "= Runge-Kutta",
  MathematicaEstimate "= Mathematica's Estimate"}]
```

Out[41]=
```
0.2 = jump gives you these estimates:

4.08149 = Euler
4.22975 = Trapezoidal
4.23993 = Midpoint
4.23653 = Runge-Kutta
4.23653 = Mathematica's Estimate
```

Now the Runge-Kutta estimate agrees with *Mathematica*'s estimate, and the trapezoidal and the midpoint estimates are beginning to shape up. Play with these by reducing the step size even more, but don't make it too small, because you don't want to paralyze your machine with huge calculations and you don't want to lose accuracy to round-off error.

Go back to the beginning of this part, change $f'[x], a$ and c and play.

Tutorials

■ T.1) Taylor's formula in reverse

T.1.a) Here is the expansion in powers of x of $f[x] = 1/(1 + x - x^2)$ through the x^{12} term:

In[1]:=
```
Clear[x,fexpan12]; f[x_] = 1/(1 - x - x^2);
expan12[x_] = Normal[Series[f[x],{x,0,12}]]
```

Out[1]=
$$1 + x + 2x^2 + 3x^3 + 5x^4 + 8x^5 + 13x^6 + 21x^7 + 34x^8 + 55x^9 + 89x^{10} + 144x^{11} + 233x^{12}$$

Use this partial expansion to come up with a table of the values of the derivatives $\{f'[0], f''[0], f'''[0], \ldots, f^{(11)}[0], f^{(12)}[0]\}$.

Answer: Taylor's formula says that the expansion of $f[x]$ in powers of $(x - b)$ is

$$f[b] + f'[b](x-b) + \frac{f''[b]}{2!}(x-b)^2 + \frac{f^{(3)}[b]}{3!}(x-b)^3 + \cdots + \frac{f^{(k)}[b]}{k!}(x-b)^k + \cdots$$

Here you are dealing with $b = 0$. So the table $\{f'[0], f''[0], f'''[0], \ldots, f^{(11)}[0], f^{(12)}[0]\}$ is:

In[2]:=
```
Clear[k]; Table[k! Coefficient[expan12[x],x^k],{k,1,12}]
```

Out[2]=
$\{1, 4, 18, 120, 960, 9360, 105840, 1370880, 19958400, 322963200,$
$5748019200, 111607372800\}$

Check:

In[3]:=
```
Table[D[f[x],{x,k}]/.x->0,{k,1,12}]
```

Out[3]=
$\{1, 4, 18, 120, 960, 9360, 105840, 1370880, 19958400, 322963200,$
$5748019200, 111607372800\}$

Mathematica had to grind on the last one, because along the way it had to do some hairy stuff.

■ T.2) Limits

T.2.a) All you know about a certain function $f[x]$ is that $f[\pi] = 0$ and $f'[\pi] = 8$.

> Calculate
> $$\lim_{x \to \pi} \frac{f[x]}{\sin[2x]}.$$

Answer: Look at:

In[4]:=
```
Clear[f,x];
Normal[Series[f[x],{x,Pi,3}]]/.{f[Pi]->0, f'[Pi]->8}
```

Out[4]=
$$8 \, (-\text{Pi} + x) + \frac{(-\text{Pi} + x)^2 \, f''[\text{Pi}]}{2} + \frac{(-\text{Pi} + x)^3 \, f^{(3)}[\text{Pi}]}{6}$$

In[5]:=
```
Normal[Series[Sin[2 x],{x,Pi,3}]]
```

Out[5]=
$$2 \, (-\text{Pi} + x) - \frac{4 \, (-\text{Pi} + x)^3}{3}$$

So,
$$\lim_{x \to \pi} \frac{f[x]}{\sin[2x]} = \frac{8}{2} = 4.$$

Not a whole lot to it.

■ T.3) Approximations and fake plots of solutions of some differential equations

Here are the approximations of a function $f[x]$ studied in the Basics:

Euler approximation of $f[x]$ (order of contact 1 with $f[x]$ at $x = b$):

In[6]:=
```
Clear[feuler,f,x,b,Derivative]
feuler[x_,b_] = f[b] + (x - b) f'[b]
```

Out[6]=
```
f[b] + (-b + x) f'[b]
```

Trapezoidal approximation of $f[x]$ (order of contact 2 with $f[x]$ at $x = b$):

In[7]:=
```
Clear[ftrap,f,x,b,Derivative]
ftrap[x_,b_] = f[b] + (x - b) (f'[b] + f'[x])/2
```

Out[7]=

$$f[b] + \frac{(-b + x)(f'[b] + f'[x])}{2}$$

Midpoint approximation of $f[x]$ (order of contact 2 with $f[x]$ at $x = b$):

In[8]:=
```
Clear[fmidpt,f,x,b,Derivative]
fmidpt[x_,b_] = f[b] + (x - b) f'[(b + x)/2]
```

Out[8]=

$$f[b] + (-b + x) f'[\frac{b + x}{2}]$$

Runge-Kutta approximation of $f[x]$ (order of contact 4 with $f[x]$ at $x = b$):

In[9]:=
```
Clear[frunge,f,x,b,Derivative]
frunge[x_,b_] = (2 fmidpt[x,b] + ftrap[x,b])/3
```

Out[9]=

Out[94]=

$$f[b] + \frac{\frac{(-b+x)(f'[b]+f'[x])}{2} + 2(f[b] + (-b+x) f'[\frac{b+x}{2}])}{3}$$

T.3.a) How do you use these approximations to fake plots of solutions of differential equations $y'[x] = F[x, y[x]]$ with given starting data $y[a]$?

Answer: Really you shouldn't. You're a lot better off using NDSolve, or other professionally written software which takes advantage of methods more advanced than these approximations. Serious research in this area is going on right now and the state of the art is changing rapidly. Nevertheless, using the four approximations above to fake the plots of solutions to differential equations gives the flavor of the more advanced methods, so it's not a bad idea to see how they can be used.

Euler approximation of $f[x]$ (order of contact 1 with $f[x]$ at $x = b$):

In[10]:=
```
Clear[feuler,f,Derivative,x,b]
feuler[x_,b_] = f[b] + (x - b) f'[b]
```

Out[10]=
```
f[b] + (-b + x) f'[b]
```

Take $(x - b) =$ jump and modify accordingly to get:

Euler update for faking the plot of $y'[x] = F[x, y[x]]$:

In[11]:=
```
Clear[nexteuler,jump,F,x,y]
nexteuler[{x_,y_},jump_] = {x,y} + jump {1,F[x,y]}
```

Out[11]=
```
{jump + x, y + jump F[x, y]}
```

Trapezoidal approximation of $f[x]$ (order of contact 2 with $f[x]$ at $x = b$):

In[12]:=
```
Clear[ftrap,f,jump,x,b]
ftrap[x_,b_] = f[b] + (x - b) (f'[b] + f'[x])/2
```

Out[12]=
$$f[b] + \frac{(-b + x)\ (f'[b] + f'[x])}{2}$$

Take $(x - b) =$ jump and modify accordingly to get:

Trapezoidal update for faking the plot of $y'[x] = F[x, y[x]]$:

In[13]:=
```
Clear[nexttrap,jump,F,x,y]
nexttrap[{x_,y_},jump_] = {x,y} +
 N[jump{1, (F[x,y] + F[x + jump,y + jump F[x,y]])/2}]
```

Out[13]=
```
{jump + x, y + 0.5 jump
  (F[x, y] + F[jump + x, y + jump F[x, y]])}
```

Most advanced folks call this a two-stage Runge-Kutta update.

Midpoint approximation of $f[x]$ (order of contact 2 with $f[x]$ at $x = b$):

In[14]:=
```
Clear[fmidpt,f,x,b]
fmidpt[x_,b_] = f[b] + (x - b) f'[(b + x)/2]
```

Out[14]=
$$f[b] + (-b + x) \, f'[\frac{b + x}{2}]$$

Take $(x - b) =$ jump and modify accordingly to get:

Midpoint update for faking the plot of $y'[x] = F[x, y[x]]$:

In[15]:=
```
Clear[nextmidpt,F,x,y,jump]
nextmidpt[{x_,y_},jump_] = {x,y} + N[jump{1,F[x + jump/2,
   y + (jump F[x + jump,y + jump F[x,y]])/2]}]
```

Out[15]=
```
{jump + x, y + jump F[0.5 jump + x,
  y + 0.5 jump F[jump + x, y + jump F[x, y]]]}
```

Most advanced folks call this a two-stage Runge-Kutta update as well.

Runge-Kutta approximation of $f[x]$ (order of contact 4 with $f[x]$ at $x = b$):

In[16]:=
```
Clear[frunge,f,x,b]
frunge[x_,b_] = (2 fmidpt[x,b] + ftrap[x,b])/3;
```

Runge-Kutta update for faking the plot of $y'[x] = F[x, y[x]]$:

In[17]:=
```
Clear[nextrunge,F,x,y,jump]
nextrunge[{x_,y_},jump_] = ExpandAll[
  (2 nextmidpt[{x,y},jump] + nexttrap[{x,y},jump])/3]
```

Out[17]=
$$\{jump + x, \, y + 0.166667 \text{ jump } F[x, y] +$$
$$\frac{2 \text{ jump } F[0.5 \text{ jump } + x, \, y + 0.5 \text{ jump } F[jump + x, \, y + jump \, F[x, y]]]}{3} +$$
$$0.166667 \text{ jump } F[jump + x, \, y + jump \, F[x, y]]\}$$

Most advanced folks call this a three-stage Runge-Kutta update.

Now try these four updates on the differential equation

$$y'[x] = \sin[x] \, y[x] \quad \text{with} \quad y[0] = 1.$$

First, see what *Mathematica*'s NDSolve does:

In[18]:=
```
a = 0; c = 4; starter = 1;
Clear[solution,x,y,fakey]
```

```
solution = NDSolve[{y'[x] == Sin[x] y[x],y[0] == starter},y[x],{x,a,c}];
fakey[x_] = y[x]/.solution[[1]]
```

Out[18]=
```
InterpolatingFunction[{0., 4.}, <>][x]
```

In[19]:=
```
Mathematicaplot = Plot[fakey[x],{x,a,c},
  PlotStyle->{{Blue,Thickness[0.02]}},
  AxesLabel->{"x","y[x]"},AxesOrigin->{a,starter},
  PlotRange->All];
```

Activate the updates:

In[20]:=
```
Clear[nexteuler,F,x,y,jump]; Clear[nexttrap]
Clear[nextmidpt,F,x,y]; Clear[nextrunge,F,x,y]
nexteuler[{x_,y_},jump_] := {x,y} + N[jump{1,F[x,y]}]
nexttrap[{x_,y_},jump_] := {x,y}+N[jump{1,(F[x,y] + F[x + jump,y + jump F[x,y]])/2}]
nextmidpt[{x_,y_},jump_] := {x,y} + N[jump{1,F[x + jump/2,y +
  (jump F[x + jump,y + jump F[x,y]])/2}]
nextrunge[{x_,y_},jump_] := (2 nextmidpt[{x,y},jump] + nexttrap[{x,y},jump])/3
```

The differential equation is

$$y'[x] = \sin[x]\, y[x] \qquad \text{with} \qquad y[0] = 1.$$

In[21]:=
```
a = 0; c = 4; starter = 1;
Clear[x,y,F]; beginner = {a,starter};
F[x_,y_] = Sin[x] y
```

Out[21]=
```
y Sin[x]
```

Here's what you get from the Euler update with jump $= (c - a)/10$:

In[22]:=
```
Clear[jump,eulerpoint,k]; Clear[Eulerplot,jump]
eulerpoint[0,jump_] = beginner;
eulerpoint[k_,jump_] := eulerpoint[k,jump] = N[nexteuler[eulerpoint[k-1,jump],jump]]
Eulerplot[jump_] := Show[Graphics[{Thickness[0.01],Red,
  Line[Table[eulerpoint[k,jump],{k,0,(c - a)/jump}]]}],
  DisplayFunction->Identity];
```

In[23]:=
```
Show[Mathematicaplot,
  Eulerplot[(c - a)/10],
  PlotRange->{-1,8},PlotLabel->Euler,
  DisplayFunction->$DisplayFunction];
```

The *Mathematica* plot of the solution is shown for reference. Here's what you get from the trapezoidal update with jump $= (c-a)/10$:

In[24]:=
```
Clear[jump,trappoint,k]; Clear[trapplot]
trappoint[0,jump_] = beginner;
trappoint[k_,jump_] := trappoint[k,jump] =
  N[nexttrap[trappoint[k-1,jump],jump]];
trapplot[jump_] := Show[Graphics[
  {Thickness[0.01],Red,Line[Table[trappoint[k,jump],
  {k,0,(c - a)/jump}]]}],Axes->True,
  AxesOrigin->beginner,DisplayFunction->Identity];
Show[Mathematicaplot,trapplot[(c - a)/10],
  PlotLabel->"trapezoidal",PlotRange->{-1,8},
  DisplayFunction->$DisplayFunction];
```

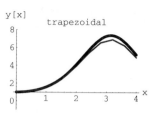

Here's what you get from the midpoint update with jump $= (c-a)/10$:

In[25]:=
```
Clear[jump,midptpoint,k]; Clear[midptplot]
midptpoint[0,jump_] = beginner;
midptpoint[k_,jump_] := midptpoint[k,jump] =
  N[nextmidpt[midptpoint[k-1,jump],jump]];
midptplot[jump_] := Show[Graphics[{Thickness[0.01],
  Red,Line[Table[midptpoint[k,jump],
  {k,0,(c - a)/jump}]]}],DisplayFunction->Identity];
Show[Mathematicaplot,midptplot[(c - a)/10],
  PlotLabel->"midpoint",PlotRange->{-1,8},
  DisplayFunction->$DisplayFunction];
```

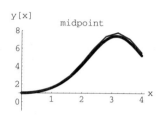

Here is what you get from the Runge-Kutta update with jump $= (c-a)/10$:

In[26]:=
```
Clear[jump,rungepoint,k]; Clear[rungeplot]
rungepoint[0,jump_] = beginner;
rungepoint[k_,jump_] := rungepoint[k,jump] =
  N[nextrunge[rungepoint[k-1,jump],jump]];
rungeplot[jump_] := Show[Graphics[
  {Thickness[0.01],Red,Line[Table[rungepoint[
  k,jump],{k,0,(c - a)/jump}]]}],
  DisplayFunction->Identity];
Show[Mathematicaplot,rungeplot[(c - a)/10],
  PlotRange->{-1,8},PlotLabel->"Runge-Kutta",
  DisplayFunction->$DisplayFunction];
```

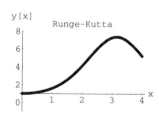

Very, very good.

Now see what you get when you reduce the jump from

In[27]:=
```
N[(c - a)/10]
```
Out[27]=
```
0.4
```

to

In[28]:=
```
N[(c - a)/20]
```
Out[28]=
```
0.2
```

In[29]:=
```
Show[Mathematicaplot,
 Eulerplot[(c - a)/20],
 PlotRange->{-1,8},
 PlotLabel->Euler,
 DisplayFunction->$DisplayFunction];
```

In[30]:=
```
Show[Mathematicaplot,
 trapplot[(c - a)/20],
 PlotLabel->"trapezoidal",
 PlotRange->{-1,8},
 DisplayFunction->$DisplayFunction];
```

In[31]:=
```
Show[Mathematicaplot,
 midptplot[(c - a)/20],
 PlotLabel->"midpoint",
 PlotRange->{-1,8},
 DisplayFunction->$DisplayFunction];
```

In[32]:=
```
Show[Mathematicaplot,
 rungeplot[(c - a)/20],
 PlotRange->{-1,8},
 PlotLabel->"Runge-Kutta",
 DisplayFunction->$DisplayFunction];
```

The plots speak for themselves.

Give It a Try

Experience with the starred ($\star$) problems will be especially beneficial for understanding later lessons.

Lesson 4.04 Taylor's Formula

■ G.1) Taylor's formula in reverse★

G.1.a) Here is the expansion in powers of x of $f[x] = 1/(1-x^2)$ through the x^{20} term:

In[1]:=
```
Clear[x,fexpan20]
f[x_] = 1/(1 - x^2);
expan20[x_] = Normal[Series[f[x],{x,0,20}]]
```

Out[1]=
$$1 + x^2 + x^4 + x^6 + x^8 + x^{10} + x^{12} + x^{14} + x^{16} + x^{18} + x^{20}$$

> Use this partial expansion in powers of x to come up with a table of the values of the derivatives $\{f'[0], f''[0], f'''[0], \ldots, f^{(19)}[0], f^{(20)}[0]\}$.

G.1.b) Here is the expansion in powers of $(x-b)$ of $f[x] = x^2 e^{-x}$ through the x^6 term:

In[2]:=
```
Clear[x,f,b,expan6]; f[x_] = x^2 E^(-x);
expan6[x_] = Normal[Series[f[x],{x,b,6}]]
```

Out[2]=
$$\frac{b^2}{E^b} + \left(\frac{2b}{E^b} - \frac{b^2}{E^b}\right)(-b+x) + \left(E^{-b} - \frac{2b}{E^b} + \frac{b^2}{2E^b}\right)(-b+x)^2 +$$
$$\left(-E^{-b} + \frac{b}{E^b} - \frac{b^2}{6E^b}\right)(-b+x)^3 + \left(\frac{1}{2E^b} - \frac{b}{3E^b} + \frac{b^2}{24E^b}\right)(-b+x)^4 +$$
$$\left(\frac{-1}{6E^b} + \frac{b}{12E^b} - \frac{b^2}{120E^b}\right)(-b+x)^5 + \left(\frac{1}{24E^b} - \frac{b}{60E^b} + \frac{b^2}{720E^b}\right)(-b+x)^6$$

> Use this partial expansion in powers of $(x-b)$ to come up with a table of the values of the derivatives $\{f'[b], f''[b], f'''[b], f^{(4)}[x], f^{(5)}[b], f^{(6)}[b]\}$.

■ G.2) Limits, Taylor's formula, and L'Hospital's rule★

G.2.a.i) All you know about a certain function $f[x]$ is that $f[3] = 0$ and $f'[3] = 9$.

> Calculate
> $$\lim_{x \to 3} \frac{x^3 - 3^3}{f[x]}.$$

G.2.a.ii) All you know about a certain function $f[x]$ is that $f'[x] = \cos[f[x]] + 4x$ and $f[1] = 0$.

> Calculate
> $$\lim_{x \to 1} \frac{f[x]}{\cos[\pi x/2]}.$$

G.2.b.i) Take two functions $f[x]$ and $g[x]$ such that

→ $f[b] = 0$ and $g[b] = 0$ and

→ $g'[b] \neq 0$

and look at:

In[3]:=
```
Clear[f,g,x,b]; Normal[Series[f[x],{x,b,2}]]/.f[b]->0
```
Out[3]=
$$(-b + x)\, f'[b] + \frac{(-b + x)^2\, f''[b]}{2}$$

In[4]:=
```
Normal[Series[g[x],{x,b,2}]]/.g[b]->0
```
Out[4]=
$$(-b + x)\, g'[b] + \frac{(-b + x)^2\, g''[b]}{2}$$

> How does this tell you that you can read off
> $$\lim_{x \to b} \frac{f[x]}{g[x]} = \frac{f'[b]}{g'[b]}?$$

Some folks call this formula L'Hospital's Rule.

G.2.b.ii) Take two functions $f[x]$ and $g[x]$ such that

→ $f[b] = 0$ and $g[b] = 0$ and

→ $f'[b] = 0$ and $g'[b] = 0$ and

→ $g''[b] \neq 0$

and look at:

In[5]:=
```
Clear[f,g,x,b]
Normal[Series[f[x],{x,b,3}]]/.{f[b]->0,f'[b]->0}
```

Out[5]=

$$\frac{(-b + x)^2 \, f''[b]}{2} + \frac{(-b + x)^3 \, f^{(3)}[b]}{6}$$

In[6]:=
```
Normal[Series[g[x],{x,b,3}]]/.{g[b]->0,g'[b]->0}
```

Out[6]=

$$\frac{(-b + x)^2 \, g''[b]}{2} + \frac{(-b + x)^3 \, g^{(3)}[b]}{6}$$

> How does this tell you that you can read off
> $$\lim_{x \to b} \frac{f[x]}{g[x]} = \frac{f''[b]}{g''[b]}?$$

Some folks call this formula L'Hospital's Rule.

G.2.b.iii) Take two functions $f[x]$ and $g[x]$ such that

→ $f[b] = 0$ and $g[b] = 0$;
→ $f'[b] = 0$ and $g'[b] = 0$;
→ $f''[b] = 0$ and $g''[b] = 0$ and
→ $g'''[b] \neq 0$

> Give a formula for
> $$\lim_{x \to b} \frac{f[x]}{g[x]}$$
> and explain where you got it.

Some folks call this formula L'Hospital's Rule.

G.2.b.iv) Professor Lipman Bers, of Columbia University and past president of the American Mathematical Society, once said that L'Hospital's rule impressed the early practitioners of calculus, but the importance of L'Hospital's rule in contemporary mathematics is not overwhelming.

> Why is L'Hospital's rule no big deal if you know Taylor's formula?

■ G.3) Taylor's formula and expansions of derivatives*

G.3.a) Here is the beginning of the expansion of $f[x] = 1/(1-x)$ in powers of x:

In[7]:=
```
Clear[x,f,beginning]; f[x_] = 1/(1 - x);
beginning[x_] = Normal[Series[f[x],{x,0,10}]]
```

Out[7]=
$$1 + x + x^2 + x^3 + x^4 + x^5 + x^6 + x^7 + x^8 + x^9 + x^{10}$$

Now look at

In[8]:=
```
beginning'[x]
```

Out[8]=
$$1 + 2x + 3x^2 + 4x^3 + 5x^4 + 6x^5 + 7x^6 + 8x^7 + 9x^8 + 10x^9$$

And

In[9]:=
```
Normal[Series[f'[x],{x,0,9}]]
```

Out[9]=
$$1 + 2x + 3x^2 + 4x^3 + 5x^4 + 6x^5 + 7x^6 + 8x^7 + 9x^8 + 10x^9$$

> What is the relationship between the expansion of $f[x] = 1/(1-x)$ in powers of x and the expansion of $f'[x]$ in powers of x?

G.3.b) Here is the beginning of the expansion of $f[x] = e^{2x}$ in powers of x:

In[10]:=
```
Clear[x,f,beginning]; f[x_] = E^(2 x);
beginning[x_] = Normal[Series[f[x],{x,0,13}]]
```

Out[10]=
$$1 + 2x + 2x^2 + \frac{4x^3}{3} + \frac{2x^4}{3} + \frac{4x^5}{15} + \frac{4x^6}{45} + \frac{8x^7}{315} + \frac{2x^8}{315} + \frac{4x^9}{2835} + \frac{4x^{10}}{14175} + \frac{8x^{11}}{155925} + \frac{4x^{12}}{467775} + \frac{8x^{13}}{6081075}$$

Now look at:

In[11]:=
```
beginning'[x]
```

Out[11]=
$$2 + 4x + 4x^2 + \frac{8x^3}{3} + \frac{4x^4}{3} + \frac{8x^5}{15} + \frac{8x^6}{45} + \frac{16x^7}{315} +$$

$$\frac{4x^8}{315} + \frac{8x^9}{2835} + \frac{8x^{10}}{14175} + \frac{16x^{11}}{155925} + \frac{8x^{12}}{467775}$$

And:

In[12]:=
```
Normal[Series[f'[x],{x,0,12}]]
```

Out[12]=

$$2 + 4x + 4x^2 + \frac{8x^3}{3} + \frac{4x^4}{3} + \frac{8x^5}{15} + \frac{8x^6}{45} + \frac{16x^7}{315} +$$
$$\frac{4x^8}{315} + \frac{8x^9}{2835} + \frac{8x^{10}}{14175} + \frac{16x^{11}}{155925} + \frac{8x^{12}}{467775}$$

> What is the relationship between the expansion of $f[x] = e^{2x}$ in powers of x and the expansion of $f'[x]$ in powers of x?

G.3.c) Look at the beginning of the expansion of a cleared function $f[x]$ in powers of $(x - b)$:

In[13]:=
```
Clear[x,b,f,beginning]
beginning[x_] = Normal[Series[f[x],{x,b,6}]]
```

Out[13]=

$$f[b] + (-b + x)\, f'[b] + \frac{(-b+x)^2\, f''[b]}{2} + \frac{(-b+x)^3\, f^{(3)}[b]}{6} +$$
$$\frac{(-b+x)^4\, f^{(4)}[b]}{24} + \frac{(-b+x)^5\, f^{(5)}[b]}{120} + \frac{(-b+x)^6\, f^{(6)}[b]}{720}$$

Now look at:

In[14]:=
```
D[beginning[x],x]
```

Out[14]=

$$f'[b] + (-b+x)\, f''[b] + \frac{(-b+x)^2\, f^{(3)}[b]}{2} + \frac{(-b+x)^3\, f^{(4)}[b]}{6} +$$
$$\frac{(-b+x)^4\, f^{(5)}[b]}{24} + \frac{(-b+x)^5\, f^{(6)}[b]}{120}$$

And:

In[15]:=
```
Normal[Series[f'[x],{x,b,5}]]
```

Out[15]=

$$f'[b] + (-b + x)\, f''[b] + \frac{(-b + x)^2\, f^{(3)}[b]}{2} + \frac{(-b + x)^3\, f^{(4)}[b]}{6} +$$
$$\frac{(-b + x)^4\, f^{(5)}[b]}{24} + \frac{(-b + x)^5\, f^{(6)}[b]}{120}$$

> For any old function $f[x]$, what is the relationship between the expansion of $f[x]$ in powers of $(x - b)$ and the expansion of $f'[x]$ in powers of $(x - b)$?

■ G.4) Pulling the expansions of $1/(1 - x)$, $\sin[x]$, $\cos[x]$ and e^x in powers of $(x - b)$ out of your back pocket

This problem appears only in the electronic version.

■ G.5) Rectangles, trapezoids, midpoints, and parabolas

G.5.a.i) Euler approximation of $f[x]$

In[16]:=
```
Clear[feuler,f,x,b,Derivative]
feuler[x_,b_] = f[b] + (x - b) f'[b];
```

Here is a plot $f'[x]$ for $f[x] = x \log[x] - x$ together with the plot of a certain rectangle:

In[17]:=
```
a = 2; c = 8; f[x_] = x Log[x] - x;
rectangle = Graphics[{Goldenrod,Thickness[0.01],
 Polygon[{{a,0},{c,0},{c,f'[a]},{a,f'[a]}}]}];
fprimeplot = Plot[f'[x],{x,a,c},
 PlotStyle->Thickness[0.01],
 DisplayFunction->Identity];
Show[rectangle,fprimeplot,
 Axes->True,AxesOrigin->{a,0},
 AxesLabel->{"x","f'[x]"},
 DisplayFunction->$DisplayFunction];
```

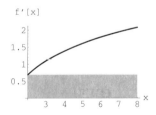

> True or false:
>
> If you take $b = a$, then
>
> $$\text{feuler}[c, b] - \text{feuler}[a, b]$$
>
> measures the area inside the rectangle.

Lesson 4.04 Taylor's Formula

G.5.a.ii) Does this work for other functions and other choices of a and c as well?

G.5.b.i) Trapezoidal approximation of $f[x]$

In[18]:=
```
Clear[ftrap,f,x,b,Derivative]
ftrap[x_,b_] = f[b] + (x - b) (f'[b] + f'[x])/2;
```

Here is a plot $f'[x]$ for $f[x] = x \log[x] - x$ together with the plot of a certain trapezoid:

In[19]:=
```
a = 2; c = 8; f[x_] = x Log[x] - x;
trapezoid = Graphics[{Goldenrod,Thickness[0.01],
    Polygon[{{a,0},{c,0},{c,f'[c]},{a,f'[a]}}]}];
fprimeplot = Plot[f'[x],{x,a,c},
    PlotStyle->Thickness[0.01],DisplayFunction->Identity];
Show[trapezoid,fprimeplot,Axes->True,AxesOrigin->{a,0},
    AxesLabel->{"x","f'[x]"},
    DisplayFunction->$DisplayFunction];
```

True or false:

If you take $b = a$, then

$$\text{ftrap}[c, b] - \text{ftrap}[a, b]$$

measures the area inside the trapezoid.

G.5.b.ii) Does this work for other functions and other choices of a and c as well?

G.5.c.i) Midpoint approximation of $f[x]$

In[20]:=
```
Clear[fmidpt,f,x,b,Derivative]
fmidpt[x_,b_] = f[b] + (x - b) f'[(b + x)/2];
```

Here is a plot of $f'[x]$ for $f[x] = x \log[x] - x$ together with the plot of a certain rectangle:

In[21]:=
```
a = 2; c = 8; f[x_] = x Log[x] - x;
trapezoid = Graphics[{Goldenrod,Thickness[0.01],
    Polygon[{{a,0},{c,0},{c,f'[(a + c)/2]},
    {a,f'[(a + c)/2]}}]}];
fprimeplot = Plot[f'[x],{x,a,c},
    PlotStyle->Thickness[0.01],DisplayFunction->Identity];
Show[trapezoid,fprimeplot,Axes->True,AxesOrigin->{a,0},
    AxesLabel->{"x","f'[x]"},
    DisplayFunction->$DisplayFunction];
```

> True or false:
>
> If you take $b = a$, then
>
> $$\text{fmidpt}[c, b] - \text{fmidpoint}[a, b]$$
>
> measures the area inside the rectangle.

G.5.c.ii) Does this work for other functions and other choices of a and c as well?

G.5.d.i) Runge-Kutta approximation of $f[x]$

In[22]:=
```
Clear[ftrap,fmidpt,frunge,f,x,b,Derivative]
ftrap[x_,b_] = f[b] + (x - b) (f'[b] + f'[x])/2;
fmidpt[x_,b_] = f[b] + (x - b) f'[(b + x)/2];
frunge[x_,b_] = (2 fmidpt[x,b] + ftrap[x,b])/3
```

Out[22]=

$$f[b] + \frac{\frac{(-b + x)(f'[b] + f'[x])}{2} + 2\left(f[b] + (-b + x)f'\left[\frac{b + x}{2}\right]\right)}{3}$$

Here is a plot of $f'[x]$ for $f[x] = x \log[x] - x$ together with the plot of a certain parabola:

In[23]:=
```
Clear[parabola]
a = 2; c = 8; f[x_] = x Log[x] - x;
parabola[x_] = Fit[{{a,f'[a]},{(a + c)/2,f'[(a + c)/2]},{c,f'[c]}},{1,x,x^2},x]
```

Out[23]=

$$-0.165651 + 0.478986\, x - 0.0247937\, x^2$$

In[24]:=
```
fprimeplot = Plot[f'[x],{x,a,c},
PlotStyle->Thickness[0.01],DisplayFunction->Identity];
parabolaplot = Plot[parabola[x],{x,a,c},
PlotStyle->{{Thickness[0.01],Red}},
DisplayFunction->Identity];
contactplot = {Graphics[{PointSize[0.03],
Point[{a,f'[a]}]}],Graphics[{PointSize[0.03],
Point[{(a + c)/2,f'[(a + c)/2]}]}],
Graphics[{PointSize[0.03],Point[{c,f'[c]}]}]};
Show[contactplot,parabolaplot,fprimeplot,
Axes->True,AxesOrigin->{a,0},AxesLabel->{"x","f'[x]"},
PlotRange->{0,f'[c]},DisplayFunction->$DisplayFunction];
```

> True or false:
> If you take $b = a$, then
> $$\text{frunge}[c, b] - \text{frunge}[a, b] = \int_a^c \text{parabola}[x]\, dx.$$
> True or false: If you take $b = a$, then
> $$\text{frunge}[t, b] - \text{frunge}[a, b] = \int_a^t \text{parabola}[x]\, dx$$
> for any t with $a \leq t \leq c$.

G.5.d.ii) Does this work for other functions and other choices of a and c as well?

G.5.d.iii) Simpson's rule

For your information: When you fit with parabolas as above, you are approximating $\int_a^c f'[x]\,dx$ by something most old timers call "Simpson's rule." Simpson's rule will give you exactly the same results you get when you estimate integrals by calculating

$$\text{Sum}[\text{frunge}[b + \text{jump}, b] - \text{frunge}[b, b]], \{b, a, c - \text{jump}, \text{jump}\}].$$

■ G.6) The midpoint and Runge-Kutta approximations versus expansions

Calculus&Mathematica is very pleased to acknowledge that the presentation here is greatly influenced by the book *Numerical Methods and Software* by David Kahaner, Cleve Moler, and Stephen Nash, Prentice-Hall, 1989. This book is a winner.

Euler approximation: Here is the Euler approximation of a function $f[x]$ at a point $x = b$:

In[25]:=
```
Clear[feuler,f,x,b,Derivative]
feuler[x_,b_] = f[b] + (x - b) f'[b];
```

The Euler approximation is what you get when you approximate $f[x]$ by the first two terms of the expansion of $f[x]$ in powers of $(x - b)$.

In[26]:=
```
{feuler[x,b],Normal[Series[f[x],{x,b,1}]]}
```
Out[26]=
```
{f[b] + (-b + x) f'[b], f[b] + (-b + x) f'[b]}
```

Midpoint approximation: Here is the midpoint approximation of a function $f[x]$ at a point $x = b$:

In[27]:=
```
Clear[fmidpt,f,x,b,Deriavtive]
fmidpt[x_,b_] = f[b] + (x - b) f'[(b + x)/2];
```

fmidpt$[x, b]$ has order of contact 2 with $f[x]$ at $x = b$:

In[28]:=
```
{Series[fmidpt[x,b],{x,b,3}],Series[f[x],{x,b,3}]}
```

Out[28]=

$$\{f[b] + f'[b](-b+x) + \frac{f''[b](-b+x)^2}{2} + \frac{f^{(3)}[b](-b+x)^3}{8} + O[-b+x]^4,$$

$$f[b] + f'[b](-b+x) + \frac{f''[b](-b+x)^2}{2} + \frac{f^{(3)}[b](-b+x)^3}{6} + O[-b+x]^4\}$$

But fmidpt$[x, b]$ is not the same as the second degree polynomial with order of contact 2 with $f[x]$ at $x = b$:

In[29]:=
```
Clear[expan2]
expan2[x_,b_] = Normal[Series[f[x],{x,b,2}]]
```

Out[29]=

$$f[b] + (-b+x) f'[b] + \frac{(-b+x)^2 f''[b]}{2}$$

In[30]:=
```
fmidpt[x,b]
```

Out[30]=

$$f[b] + (-b+x) f'[\frac{b+x}{2}]$$

Here is a plot on the interval $[b-2, b+2]$ of $f[x]$, fmidpt$[x, b]$ and expan2$[x, b]$ for $f[x] = \cos[x]$ and $b = 0$:

In[31]:=
```
f[x_] = Cos[x]; b = 0;
Plot[{f[x],fmidpt[x,b],expan2[x,b]},
 {x,b - 2,b + 2},
 PlotStyle->{{Thickness[0.02],Blue},
  {Thickness[0.01],Red},
  {Thickness[0.005],Brown}},
 AxesLabel->{"x",""}];
```

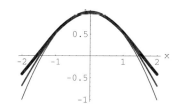

Even though fmidpt$[x, b]$, expan2$[x, b]$ and $f[x]$ have order of contact 2 at $x = b$, the quality of the approximation of $f[x]$ by fmidpt$[x, b]$ is not the same as the quality of the approximation of $f[x]$ by expan2$[x, b]$.

G.6.a.i) Which is better in the situation plotted above?

G.6.a.ii) Try experiments like the above for functions like $f[x] = e^x$ with $b = 1$, $f[x] = \log[x]$ with $b = 3$ and some others of your own choice. Discuss what you see.

G.6.a.iii) Which do you expect to give the better approximation of any old function $f[x]$ on an interval $[b-2, b+2]$:

$$\text{expan2}[x, b] = f[b] + f'[b](x - b) + \frac{f''[b]}{2}(x - b)$$

or

$$\text{fmidpt}[x, b] = f[b] + f'\left[\frac{x+b}{2}\right](x - b)?$$

G.6.b) The Runge-Kutta approximation of a function $f[x]$ at a point $x = b$ is a weighted average of the midpoint and the trapezoidal approximations:

In[32]:=
```
Clear[fmidpt,ftrap,frunge,f,x,b]
ftrap[x_,b_] = f[b] + (x - b) (f'[x] + f'[b])/2;
fmidpt[x_,b_] = f[b] + (x - b) f'[(x + b)/2];
frunge[x_,b_] = (2 fmidpt[x,b] + ftrap[x,b])/3
```

Out[32]=

$$f[b] + \frac{\dfrac{(-b + x)(f'[b] + f'[x])}{2} + 2\left(f[b] + (-b + x)\, f'\!\left[\dfrac{b + x}{2}\right]\right)}{3}$$

$\text{frunge}[x, b]$ has order of contact 4 with $f[x]$ at $x = b$.

In[33]:=
```
{Series[frunge[x,b],{x,b,5}],Series[f[x],{x,b,5}]}
```

Out[33]=

$$\Big\{ f[b] + f'[b](-b+x) + \frac{f''[b](-b+x)^2}{2} + \frac{f^{(3)}[b](-b+x)^3}{6} + \frac{f^{(4)}[b](-b+x)^4}{24} + \frac{5\, f^{(5)}[b](-b+x)^5}{576} + O[-b+x]^6,$$

$$f[b] + f'[b](-b+x) + \frac{f''[b](-b+x)^2}{2} + \frac{f^{(3)}[b](-b+x)^3}{6} + \frac{f^{(4)}[b](-b+x)^4}{24} + \frac{f^{(5)}[b](-b+x)^5}{120} + O[-b+x]^6 \Big\}$$

But $\text{frunge}[x, b]$ is not the same as the fourth degree polynomial with order of contact 4 with $f[x]$ at $x = b$:

In[34]:=
```
Clear[expan4]
expan4[x_,b_] = Normal[Series[f[x],{x,b,4}]]
```
Out[34]=

$$f[b] + (-b + x)\, f'[b] + \frac{(-b+x)^2\, f''[b]}{2} + \frac{(-b+x)^3\, f^{(3)}[b]}{6} + \frac{(-b+x)^4\, f^{(4)}[b]}{24}$$

In[35]:=
```
frunge[x,b]
```
Out[35]=

$$\frac{f[b] + \dfrac{(-b+x)\,(f'[b] + f'[x])}{2} + 2\left(f[b] + (-b+x)\, f'\!\left[\dfrac{b+x}{2}\right]\right)}{3}$$

Here is a plot on the interval $[b-3, b+3]$ of $f[x]$, frunge$[x,b]$ and expan4$[x,b]$ for $f[x] = \cos[x]$ and $b = 0$:

In[36]:=
```
f[x_] = Cos[x]; b = 0;
Plot[{f[x],frunge[x,b],expan4[x,b]},
  {x,b - 3,b + 3},
  PlotStyle->{{Thickness[0.02],Blue},
    {Thickness[0.01],Red},
    {Thickness[0.005],Brown}},
  AxesLabel->{"x",""}];
```

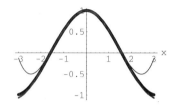

Even though frunge$[x,b]$ and expan4$[x,b]$ have order of contact 4 with $f[x]$ at $x = b$, the quality of the approximation of $f[x]$ by frunge$[x,b]$ is not the same as the quality of the approximation of $f[x]$ by expan4$[x,b]$.

G.6.b.i) Which is better in the situation plotted above?

G.6.b.ii) Try experiments like the above for functions like $f[x] = e^x$ with $b = 1$, $f[x] = \log[x]$ with $b = 3$ and some others of your own choice. Discuss what you see.

G.6.b.iii) Which do you expect to give the better approximation of a function $f[x]$ on an interval $[b-3, b+3]$:

$$\text{expan4}[x,b] = f[b] + f'[b]\,(x-b) + \left(\frac{f''[b]}{2}\right)(x-b)^2$$
$$+ \left(\frac{f'''[b]}{3!}\right)(x-b)^3 + \left(\frac{f''''[b]}{4!}\right)(x-b)^4$$

or frunge$[x,b]$?

134 Lesson 4.04 Taylor's Formula

G.6.c.i) True or false:

When you calculate fmidpt$[x,b]$ or frunge$[x,b]$, then you need calculate only $f'[x]$.

G.6.c.ii) True or false:

When you calculate expan2$[x,b]$, then you need to calculate both first and second derivatives.

G.6.c.iii) True or false:

When you calculate expan4$[x,b]$, then you need to calculate first, second, third, and fourth derivatives.

G.6.c.iv) This problem appears only in the electronic version.

■ G.7) Midpoint versus trapezoidal approximation

G.7.a) Here are the trapezoidal and the midpoint approximations of $f[x]$ near a point b:

In[37]:=
```
Clear[ftrap,f,x,b,Derivative]
ftrap[x_,b_] = f[b] + (x - b) (f'[b] + f'[x])/2
```

Out[37]=
$$f[b] + \frac{(-b + x)\,(f'[b] + f'[x])}{2}$$

In[38]:=
```
Clear[fmidpt,f,x,b]
fmidpt[x_,b_] = f[b] + (x - b) f'[(b + x)/2]
```

Out[38]=
$$f[b] + (-b + x)\,f'[\frac{b + x}{2}]$$

Check them out in the case of $f[x] = \cos[x]$ and $b = 0$:

In[39]:=
```
f[x_] = Cos[x]; b = 0;
Plot[{f[x],fmidpt[x,b],ftrap[x,b]},
 {x,b - 2,b + 2},
 PlotStyle->{{Thickness[0.02],Blue},
 {Thickness[0.01],Red},{Thickness[0.005],Brown}},
 AxesLabel->{"x",""}];
```

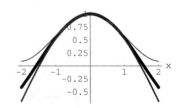

For this $f[x]$, fmidpt$[x,b]$ (thickest) is a slightly better approximation of $f[x]$ than ftrap$[x,b]$ (thinnest) even though both approximations have order of contact 2 with $f[x]$ at $x = b$. See what happens for a different function and a different b:

In[40]:=
```
Clear[f]; b = 1;
f[x_] = x + E^(x/2) Cos[3 x];
Plot[{f[x],fmidpt[x,b],ftrap[x,b]},
 {x,b - 2,b + 2},
 PlotStyle->{{Thickness[0.02],Blue},
 {Thickness[0.01],Red},{Thickness[0.005],Brown}},
 AxesLabel->{"x",""}];
```

Again, for x's near b, fmidpoint$[x,b]$ hugs the $f[x]$ curve slightly better than ftrap$[x,b]$.

> Do a couple more for functions of your own choice and report your results.

G.7.b) When Accurate Ann did these experiments, she exclaimed: "I know why they came out the way they did! And I know why it is likely to happen this way no matter what $f[x]$ and b you try. Look at this:"

In[41]:=
```
Clear[ftrap,f,x,b]
ftrap[x_,b_] = f[b] + (x - b) (f'[b] + f'[x])/2;
Clear[fmidpt,f,x,b]
fmidpt[x_,b_] = f[b] + (x - b) f'[(b + x)/2];
```

In[42]:=
```
{Normal[Series[fmidpt[x,b],{x,b,3}]],
 Normal[Series[f[x],{x,b,3}]],
 Normal[Series[ftrap[x,b],{x,b,3}]]}
```

Out[42]=

$$\{f[b] + (-b + x) f'[b] + \frac{(-b + x)^2 f''[b]}{2} + \frac{(-b + x)^3 f^{(3)}[b]}{8},$$

$$f[b] + (-b + x) f'[b] + \frac{(-b + x)^2 f''[b]}{2} + \frac{(-b + x)^3 f^{(3)}[b]}{6},$$

$$f[b] + (-b + x) f'[b] + \frac{(-b + x)^2 f''[b]}{2} + \frac{(-b + x)^3 f^{(3)}[b]}{4}\}$$

Then she said, "Look at":

In[43]:=
```
{N[1/6 - 1/8],N[1/4 - 1/6]}
```

Out[43]=
```
{0.0416667, 0.0833333}
```

And then she said, "This explains why the experiments came out the way they did."

> What was she driving at?

G.7.c) Not one to stop when she is ahead, Accurate Ann went on to say, "Look at this":

In[44]:=
```
Clear[ftrap,f,x,b]
ftrap[x_,b_] = f[b] + (x - b) (f'[b] + f'[x])/2
```

Out[44]=

$$f[b] + \frac{(-b + x)\ (f'[b] + f'[x])}{2}$$

In[45]:=
```
Clear[fmidpt,f,x,b]
fmidpt[x_,b_] = f[b] + (x - b) f'[(b + x)/2]
```

Out[45]=

$$f[b] + (-b + x)\ f'[\frac{b + x}{2}]$$

In[46]:=
```
{Normal[Series[fmidpt[x,b],{x,b,3}]],
 Normal[Series[f[x],{x,b,3}]],
 Normal[Series[ftrap[x,b],{x,b,3}]]}
```

Out[46]=

$$\{f[b] + (-b + x)\ f'[b] + \frac{(-b + x)^2\ f''[b]}{2} + \frac{(-b + x)^3\ f^{(3)}[b]}{8},$$

$$f[b] + (-b + x)\ f'[b] + \frac{(-b + x)^2\ f''[b]}{2} + \frac{(-b + x)^3\ f^{(3)}[b]}{6},$$

$$f[b] + (-b + x)\ f'[b] + \frac{(-b + x)^2\ f''[b]}{2} + \frac{(-b + x)^3\ f^{(3)}[b]}{4}\}$$

And then she said: "If the third derivative $f^{(3)}[b]$ is positive, then:

→ for x slightly to the right of b, you expect to have

$\text{fmidpt}[x, b] < f[x] < \text{ftrap}[x, b]$

and

→ for x slightly to the left of b, you expect to have

$\text{ftrap}[x, b] < f[x] < \text{fmidpt}[x, b]$.

But if the third derivative $f^{(3)}[b]$ is negative, then:

→ for x slightly to the right of b, you expect to have
$$\text{ftrap}[x,b] < f[x] < \text{fmidpt}[x,b]$$
and
→ for x slightly to the left of b, you expect to have
$$\text{fmidpt}[x,b] < f[x] < \text{ftrap}[x,b].\text{"}$$

> Where did she get this idea?

G.7.d) Very satisified with herself, Accurate Ann went on to say, "Now I see why a weighted average, like frunge$[x,b]$, of ftrap$[x,b]$ and fmidpt$[x,b]$ can be expected to be a better approximation of $f[x]$ than either ftrap$[x,b]$ or fmidpt$[x,b]$."

> What was she driving at?

G.7.e) With an amazed Bubba staring at her, his mouth hanging half open, Accurate Ann seized her moment of triumph and said: "Now it's clear to me for a very small jump size, the frunge$[x,b]$ method of estimating $\int_a^c f'[x]dx$ beats the fmidpt$[x,b]$ method and the fmidpt$[x,b]$ method beats the ftrap$[x,b]$ method. And if you don't believe me, check it out for yourself."

In[47]:=
```
Clear[ftrap,f,x,b,Derivative]; Clear[fmidpt]; Clear[frunge]
ftrap[x_,b_] = f[b] + (x - b) (f'[b] + f'[x])/2;
fmidpt[x_,b_] = f[b] + (x - b) f'[(b + x)/2];
frunge[x_,b_] = (2 fmidpt[x,b] + ftrap[x,b])/3;
f'[x_] = Sin[x^3]; a = 0.0; c = 1.5;
MathematicaEstimate = NIntegrate[f'[x],{x,a,c}];
Clear[b,jump,EulerEstimate,TrapezoidalEstimate,MidpointEstimate,SimpsonEstimate]
EulerEstimate[jump_] := N[Sum[feuler[b + jump,b] - feuler[b,b],{b,a,c - jump,jump}]];
TrapezoidalEstimate[jump_] := N[Sum[
ftrap[b + jump,b] - ftrap[b,b],{b,a,c - jump,jump}]];
MidpointEstimate[jump_] := N[Sum[
fmidpt[b + jump,b] - fmidpt[b,b],{b,a,c - jump,jump}]];
RungeKuttaEstimate[jump_] := N[Expand[Sum[
frunge[b + jump,b] - frunge[b,b],{b,a,c - jump,jump}]]];
jump = (c - a)/20;
ColumnForm[{jump "= jump gives you these estimates:",
"",TrapezoidalEstimate[jump] "= Trapezoidal",
MidpointEstimate[jump] "= Midpoint",
RungeKuttaEstimate[jump] "= Runge-Kutta",
MathematicaEstimate "= MathematicaEstimate"}]
```
Out[47]=
```
0.075 = jump gives you these estimates:

0.583791 = Trapezoidal
0.588435 = Midpoint
0.586887 = Runge-Kutta
0.586883 = MathematicaEstimate
```

```
In[48]:=
    jump = (c - a)/40;
    ColumnForm[{jump "= jump gives you these estimates:",
    "",TrapezoidalEstimate[jump] "= Trapezoidal",
    MidpointEstimate[jump] "= Midpoint",
    RungeKuttaEstimate[jump] "= Runge-Kutta",
    MathematicaEstimate "= MathematicaEstimate"}]
Out[48]=
    0.0375 = jump gives you these estimates:

    0.586113 = Trapezoidal
    0.587269 = Midpoint
    0.586884 = Runge-Kutta
    0.586883 = MathematicaEstimate
```

> Where did Accurate Ann get all this insight?

■ G.8) Approximations and fake plots of solutions of differential equations

This problem appears only in the electronic version.

■ G.9) The kissing parabola

Given a function $f[x]$ and given a number b, some folks call the second degree polynomial that has order of contact 2 with $f[x]$ at $x = b$ by the name "kissing parabola at $x = b$." The formula for the kissing parabola for $f[x]$ at $x = b$ is:

```
In[49]:=
    Clear[f,b,x,kiss]
    kiss[x_,b_] = Normal[Series[f[x],{x,b,2}]]
Out[49]=
                              2
                      (-b + x)  f''[b]
    f[b] + (-b + x) f'[b] + ─────────────────
                                   2
```

G.9.a.i) Plot $f[x] = e^{-(x-1)^2}$ and its kissing parabola at $b = 1$ on the same axes for $0 \leq x \leq 2$. Describe what you see.

G.9.a.ii) Plot in true scale $f[x] = \cos[x]$ and its kissing parabola at $b = 0$, $b = \pi$, $b = 2\pi$, and $b = 3\pi$ on the same axes for $-\pi \leq x \leq 4\pi$. Use the plotting option PlotRange->{-4, 4}.

G.9.b.i) Here is a plot of $f[x] = (x^4 + 18x^3) \cos[x] e^{-x^2}$ on the interval $-3 \leq x \leq 3$.

In[50]:=
```
Clear[f,x]
f[x_] = (x^4 + 18 x^3) Cos[x] E^(-x^2);
Plot[f[x],{x,-3,3},
 PlotStyle->{{Thickness[0.01],Blue}}];
```

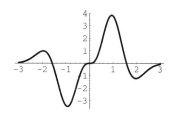

Note the high point sitting just to the left above the point $\{1, f[1]\}$. Look at a plot of $f[x]$ and its kissing parabola at $b = 1$ on the interval $1/2 \leq x \leq 3/2$.

In[51]:=
```
Clear[b,x,kiss]
kiss[x_,b_] = Normal[Series[f[x],{x,b,2}]];
Plot[{f[x],kiss[x,1]},{x,1/2,3/2},
 PlotStyle->{{Thickness[0.01],Blue},
  {Thickness[0.01],Red}}];
```

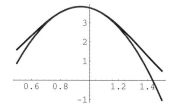

Which is closer to the maximizer of $f[x]$:

$x = 1$ or the maximizer of the kissing parabola at $a = 1$?

G.9.b.ii) First find the maximizer $x[1]$ of the kissing parabola at $b = 1$.

Then find the maximizer $x[2]$ of the kissing parabola at $b = x[1]$.

G.9.b.iii) Iterate the procedure you began in part G.9.b.i), and continue to iterate, obtaining points $x[1], x[2], x[3]$, etc. Stop when the points begin to cluster at one point. Give the value of this point. How well does this point estimate the true maximizer of $f[x]$ for $-3 \leq x \leq 3$?

LESSON 4.05

Barriers to Convergence

Basics

■ B.1) Barriers and complex numbers

B.1.a) Look at these attempts to approximate $f[x] = 1/\left(4 - 2x + x^2\right)$ by early parts of its expansion in powers of x:

In[1]:=
```
Clear[f,approx,x]; f[x_] = 1/(4 - 2 x + x^2);
approx[x_] = Normal[Series[f[x],{x,0,8}]];
Plot[{f[x],approx[x]},{x,-2.2 ,2.2},
 PlotStyle->{{Thickness[0.02],Blue},
 {Thickness[0.01],Red}},PlotRange->{-0.3,0.8},
 AxesLabel->{"x",""},PlotLabel->
 "Powers of x centered at x = 0"];
```

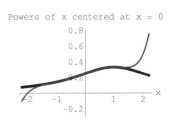

The approximation breaks down on the left and the right. Try to recover by using more of the expansion:

In[2]:=
```
Clear[approx];
approx[x_] = Normal[Series[f[x],{x,0,14}]];
Plot[{f[x],approx[x]},{x,-2.2 ,2.2},
 PlotStyle->{{Thickness[0.02],Blue},
 {Thickness[0.01],Red}},PlotRange->{-0.3,0.8},
 AxesLabel->{"x",""},PlotLabel->
 "Powers of x centered at x = 0"];
```

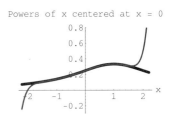

Again:

141

Lesson 4.05 Barriers to Convergence

```
In[3]:=
  Clear[approx];
  approx[x_] = Normal[Series[f[x],{x,0,18}]];
  Plot[{f[x],approx[x]},{x,-2.2 ,2.2},
  PlotStyle->{{Thickness[0.02],Blue},
  {Thickness[0.01],Red}},PlotRange->{-0.3,0.8},
  AxesLabel->{"x",""},PlotLabel->
  "Powers of x centered at x = 0"];
```

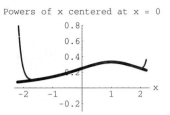

No matter what you do, you seem to run up against a barrier near $x = -2$ and $x = +2$.

Go ahead and play by using more and more of the expansion.

> What piece of advanced mathematics explains why these barriers are mathematical realities?

Answer: The key to determining the exact location of the barriers and to learning why they are not figments of your imagination comes from singularities and complex numbers.

The function is $f[x] = 1/\left(4 - 2x + x^2\right)$. The function has a complex singularity when its denominator is 0:

```
In[4]:=
  Clear[z]; Denominator[f[z]]
```

Out[4]=
$$4 - 2z + z^2$$

```
In[5]:=
  singularities = Simplify[Solve[Denominator[f[z]] == 0,z]]
```

Out[5]=
{{z -> 1 + I Sqrt[3]}, {z -> 1 - I Sqrt[3]}}

The complex singularities are $1 + i\sqrt{3}$ and $1 - i\sqrt{3}$.

Plot the points $\{1, \sqrt{3}\}$ and $\{1, -\sqrt{3}\}$ together with the center of the plotting interval $\{0,0\}$:

```
In[6]:=
  sing1 = {1,Sqrt[3]}; sing2 = {1,-Sqrt[3]};
  center = {0,0};
  points = {Graphics[{Red,PointSize[0.04],
  Point[sing1]}],Graphics[{Red,PointSize[0.04],
  Point[sing2]}],Graphics[{Blue,PointSize[0.04],
  Point[center]}]};
  labels = {Graphics[Text["sing",sing1]],
  Graphics[Text["sing",sing2]],
  Graphics[Text["center",center]]};
  Show[points,labels,Axes->True,PlotRange->All];
```

Run line segments from the center of the plotting interval to each singularity:

In[7]:=
```
lines = {Graphics[Line[{center,sing1}]],
  Graphics[Line[{center,sing2}]]};
Show[points,lines,labels,Axes->True,
  PlotRange->All];
```

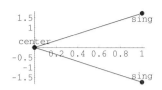

The top line connects the points:

In[8]:=
```
{center, sing1}
```
Out[8]=
```
{{0, 0}, {1, Sqrt[3]}}
```

Its length is:

In[9]:=
```
Sqrt[(1 - 0)^2 + (Sqrt[3] - 0)^2]
```
Out[9]=
```
2
```

The bottom line connects the points:

In[10]:=
```
{center, sing2}
```
Out[10]=
```
{{0, 0}, {1, -Sqrt[3]}}
```

Its length is:

In[11]:=
```
Sqrt[(1 - 0)^2 + (-Sqrt[3] - 0)^2]
```
Out[11]=
```
2
```

Both of these line segments are 2 units long. For this function,

$$R = 2$$

is the magic number.

Advanced mathematics tells you that when you expand in powers of x, then:

→ The approximations will hold up beautifully on intervals $(0 - R, 0 + R) = (-2, 2)$,

→ but the approximations will break down outside $[0 - R, 0 + R] = [-2, 2]$.

This is why you ran into barriers at $x = -R = -2$ and $x = R = 2$ above.

B.1.b) Look at these attempts to approximate $f[x] = 8/(4 + x^2)$ by early parts of its expansion in powers of $(x - 1)$:

Lesson 4.05 Barriers to Convergence

In[12]:=
```
b = 1; Clear[f,approx,x]; f[x_] = 8/(4 + x^2);
approx[x_] = Normal[Series[f[x],{x,b,8}]];
Plot[{f[x],approx[x]},{x,b - 3 ,b + 3},
PlotStyle->{{Thickness[0.02],Blue},
{Thickness[0.01],Red}},PlotRange->{-0.3,2.3},
AxesLabel->{"x",""},PlotLabel->
"Powers of (x - 1) centered at x = 1",
Epilog->{{Blue,PointSize[0.04],Point[{b,0}]},
Text["expansion point",{b,0},{0,-2}]}];
```

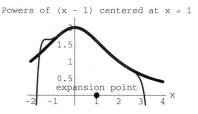

The approximation breaks down on the left and the right. Try to recover by using more of the expansion:

In[13]:=
```
Clear[approx];
approx[x_] = Normal[Series[f[x],{x,b,14}]];
Plot[{f[x],approx[x]},{x,b - 3 ,b + 3},
PlotStyle->{{Thickness[0.02],Blue},
{Thickness[0.01],Red}},PlotRange->{-0.3,2.3},
AxesLabel->{"x",""},PlotLabel->
"Powers of (x - 1) centered at x = 1",
Epilog->{{Blue,PointSize[0.04],Point[{b,0}]},
Text["expansion point",{b,0},{0,-2}]}];
```

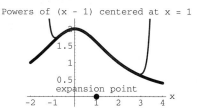

Again:

In[14]:=
```
Clear[approx];
approx[x_] = Normal[Series[f[x],{x,b,18}]];
Plot[{f[x],approx[x]},{x, b - 3 ,b + 3},
PlotStyle->{{Thickness[0.02],Blue},
{Thickness[0.01],Red}},PlotRange->{-0.3,2.3},
AxesLabel->{"x",""},PlotLabel->
"Powers of (x - 1) centered at x = 1",
Epilog->{{Blue,PointSize[0.04],Point[{b,0}]},
Text["expansion point",{b,0},{0,-2}]}];
```

No matter what you do, you seem to run up against barriers on the left and the right.

Go ahead and play by using more and more of the expansion.

> Where are the precise locations of these barriers?

Answer: The function is $f[x] = 8/\left(4 + x^2\right)$. The function has a complex singularity when its denominator is 0:

In[15]:=
```
Clear[z]; Denominator[f[z]]
```

Out[15]=
$4 + z^2$

In[16]:=
```
singularities = Simplify[Solve[Denominator[f[z]] == 0,z]]
```

Out[16]=
{{z -> 2 I}, {z -> -2 I}}

The complex singularities are $2i$ and $-2i$. Plot $\{0,2\}$ and $\{0,-2\}$ together with the center of the plotting interval $\{b,0\} = \{1,0\}$:

In[17]:=
```
sing1 = {0,2}; sing2 = {0,-2}; center = {b,0};
points = {Graphics[{Red,PointSize[0.04],
Point[sing1]}],
Graphics[{Red,PointSize[0.04],Point[sing2]}],
Graphics[{Blue,PointSize[0.04],Point[center]}]};
labels = {Graphics[Text["sing",sing1]],
Graphics[Text["sing",sing2]],
Graphics[Text["center",center]]};
Show[points,labels,Axes->True,PlotRange->All];
```

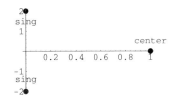

Run line segments from the center of the plotting interval to each singularity:

In[18]:=
```
lines = {Graphics[Line[{center,sing1}]],
Graphics[Line[{center,sing2}]]};
Show[points,lines,labels,
Axes->True,PlotRange->All];
```

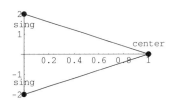

The top line connects the points:

In[19]:=
```
{center, sing1}
```

Out[19]=
{{1, 0}, {0, 2}}

Its length is:

In[20]:=
```
Sqrt[(0 - 1)^2 + (2 - 0)^2]
```

Out[20]=
Sqrt[5]

The bottom line connects the points:

In[21]:=
```
{center, sing2}
```

Lesson 4.05 Barriers to Convergence

Out[21]=
{{1, 0}, {0, -2}}

Its length is:

In[22]:=
```
Sqrt[(0 - 1)^2 + (-2 - 0)^2]
```
Out[22]=
Sqrt[5]

Both of these line segments are $\sqrt{5}$ units long. For this situation,
$$R = \sqrt{5}$$
is the magic number. When you expand $f[x]$ in powers of $(x-1)$, then

→ the approximations in powers of $(x-1)$ will hold up beautifully on intervals
$$(1 - R, 1 + R) = \left(1 - \sqrt{5}, 1 + \sqrt{5}\right),$$

→ but the approximations in powers of $(x-1)$ will break down outside
$$[1 - R, 1 + R] = [1 - \sqrt{5}, 1 + \sqrt{5}].$$

This is why you ran into barriers at $x = 1 - \sqrt{5}$ and $x = 1 + \sqrt{5}$. Take a look:

In[23]:=
```
b = 1; R = Sqrt[5];
leftbarrier = b - R; rightbarrier = b + R;
Clear[approx]; approx[x_] = Normal[Series[f[x],{x,b,18}]];
seriousplot = Plot[{f[x],approx[x]},{x, b - R ,b + R},
PlotStyle->{{Thickness[0.02],Blue},{Thickness[0.01],Red}},PlotRange->{-0.3,2.3},
Epilog->{{Blue,PointSize[0.04],Point[{b,0}]},Text["expansion point",{b,0},{0,-2}]},
DisplayFunction->Identity]; barriers = {Graphics[{Red,Line[{{leftbarrier,-0.3},
{leftbarrier,2.3}}]}],Graphics[{Red,Line[{{rightbarrier,-0.3},{rightbarrier,2.3}}]}]};
```

In[24]:=
```
Show[seriousplot,barriers,
AxesLabel->{"x",""},PlotLabel->
"Powers of (x - 1) centered at x = 1",
DisplayFunction->$DisplayFunction];
```

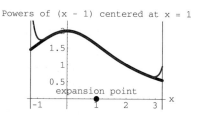

No matter how many terms of the expansion of $f[x]$ in powers of $(x-1)$ you go with, you will not be able to break the barriers at $x = 1 - \sqrt{5}$ and $x = 1 + \sqrt{5}$.

B.1.c) Given a function $f[x]$, how do you go about finding the barriers for approximating $f[x]$ with partial expansions in powers of $(x - b)$?

Basics (B.1)

Answer: You just do it as above, only you do it a bit more carefully. For instance, if you are expanding in powers of $(x-2)$ and your function is:

In[25]:=
```
Clear[f,x]
f[x_] = 29/(x^4 + x^3 + 9 x^2 + x + 28)
```

Out[25]=
$$\frac{29}{28 + x + 9 x^2 + x^3 + x^4}$$

you look at:

In[26]:=
```
f[z]
```

Out[26]=
$$\frac{29}{28 + z + 9 z^2 + z^3 + z^4}$$

In[27]:=
```
f'[z]
```

Out[27]=
$$\frac{-29 (1 + 18 z + 3 z^2 + 4 z^3)}{(28 + z + 9 z^2 + z^3 + z^4)^2}$$

In[28]:=
```
Together[f''[z]]
```

Out[28]=
$$\frac{58 (-251 - 57 z + 78 z^2 + 74 z^3 + 87 z^4 + 15 z^5 + 10 z^6)}{(28 + z + 9 z^2 + z^3 + z^4)^3}$$

It becomes clear that the only singularities of $f[z]$, or of any of its derivatives, happen when the denominator of $f[z] = 0$:

In[29]:=
```
Denominator[f[z]]
```

Out[29]=
$$28 + z + 9 z^2 + z^3 + z^4$$

In[30]:=
```
Solve[Denominator[f[z]] == 0]
```

Out[30]=
$$\{\{z \to \frac{1 + I \, \text{Sqrt}[15]}{2}\},$$
$$\{z \to \frac{1 - I \, \text{Sqrt}[15]}{2}\},$$

$$\{z \to \frac{-2 + I\ \text{Sqrt}[24]}{2}\},$$
$$\{z \to \frac{-2 - I\ \text{Sqrt}[24]}{2}\}\}$$

Four complex singularities. Because you are expanding in powers of $(x-2)$, you go with $b=2$ and plot:

In[31]:=
```
b = 2; sing1 = {1/2,Sqrt[15]/2};
sing2 = {1/2,-Sqrt[15]/2};
sing3 = {-2/2,Sqrt[24]/2};
sing4 = {-2/2,-Sqrt[24]/2};
center = {b,0}; h = 0.25; points = {Graphics[
{Red,PointSize[0.04],Point[sing1-{0,h}]}],
Graphics[{Red,PointSize[0.04],Point[sing2+{0,h}]}],
Graphics[{Red,PointSize[0.04],Point[sing3-{0,h}]}],
Graphics[{Red,PointSize[0.04],Point[sing4+{0,h}]}],
Graphics[{Blue,PointSize[0.04],Point[center]}]};
labels = {Graphics[Text["sing",sing1]],
Graphics[Text["sing",sing2]],
Graphics[Text["sing",sing3]],
Graphics[Text["sing",sing4]],
Graphics[Text["center",center]]};
Show[points,labels,Axes->True,PlotRange->All];
```

Run line segments from the center $\{b, 0\}$ of the plotting interval to each singularity:

In[32]:=
```
lines = {Graphics[Line[{center,sing1}]],
Graphics[Line[{center,sing2}]],
Graphics[Line[{center,sing3}]],
Graphics[Line[{center,sing4}]]};
Show[points,lines,labels,
Axes->True,PlotRange->All];
```

Now calculate the lengths of these lines. One line connects:

In[33]:=
```
{center,sing1}
```

Out[33]=
$$\{\{2,\ 0\},\ \{\frac{1}{2},\ \frac{\text{Sqrt}[15]}{2}\}\}$$

Its length is:

In[34]:=
```
length1 = Sqrt[(1/2 - 2)^2 + (Sqrt[15]/2 - 0)^2]
```

Out[34]=
```
Sqrt[6]
```

Basics (B.1)

Another line connects:

In[35]:=
  {center,sing2}

Out[35]=
  {{2, 0}, {$\frac{1}{2}$, $\frac{-\text{Sqrt}[15]}{2}$}}

Its length is:

In[36]:=
  length2 = Sqrt[(1/2 - 2)^2 + (-Sqrt[15]/2 - 0)^2]

Out[36]=
  Sqrt[6]

Another line connects:

In[37]:=
  {center,sing3}

Out[37]=
  {{2, 0}, {-1, Sqrt[6]}}

Its length is:

In[38]:=
  length3 = Sqrt[(-1 - 2)^2 + (Sqrt[6] - 0)^2]

Out[38]=
  Sqrt[15]

The fourth line connects:

In[39]:=
  {center,sing4}

Out[39]=
  {{2, 0}, {-1, -Sqrt[6]}}

Its length is:

In[40]:=
  length4 = Sqrt[(-1 - 2)^2 + (-Sqrt[6] - 0)^2]

Out[40]=
  Sqrt[15]

Now look at:

In[41]:=
  {length1,length2,length3,length4}

Out[41]=
  {Sqrt[6], Sqrt[6], Sqrt[15], Sqrt[15]}

Or:

In[42]:=
  N[{length1,length2,length3,length4}]

Lesson 4.05 Barriers to Convergence

Out[42]=
{2.44949, 2.44949, 3.87298, 3.87298}

The smallest of these numbers is $\sqrt{6}$. $R = \sqrt{6}$ is the magic number for this situation. This tells you that the expansion of $f[x]$ in powers of $(x - 2)$ will do a good job inside

$$(2 - R, 2 + R) = \left(2 - \sqrt{6}, 2 + \sqrt{6}\right).$$

This also says that the expansion of $f[x]$ in powers of $(x - 2)$ will freak out outside the interval

$$[2 - R, 2 + R] = [2 - \sqrt{6}, 2 + \sqrt{6}].$$

Why? Because the barriers are at $x = 2 - R = 2 - \sqrt{6}$ and $x = 2 + R = 2 + \sqrt{6}$. See it now:

In[43]:=
```
b = 2; R = Sqrt[6];
leftbarrier = b - R; rightbarrier = b + R;
Clear[approx]; approx[x_] = Normal[Series[f[x],{x,b,18}]];
seriousplot = Plot[{f[x],approx[x]},{x, b - R ,b + R},PlotStyle->
 {{Thickness[0.02],Blue},{Thickness[0.01],Red}},PlotRange->{-0.3,2.3},
 Epilog->{{Blue,PointSize[0.04],Point[{b,0}]},Text["expansion point",{b,0},{0,-2}]},
 DisplayFunction->Identity];
barriers = {Graphics[{Red,Line[{{leftbarrier,-0.3},{leftbarrier,1.5}}]}],
 {Graphics[{Red,Line[{{rightbarrier,-0.3},{rightbarrier,1.5}}]}]}};
```

In[44]:=
```
Show[seriousplot,barriers,
AxesLabel->{"x",""},
PlotLabel->
"Powers of (x - 2) centered at x = 2",
PlotRange->{-0.3,1.5},
DisplayFunction->$DisplayFunction];
```

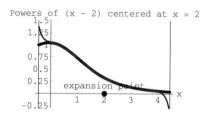

Uh-huh. You've got the right one, baby.

B.1.d) Where do you go to obtain a full understanding of the mathematics that links barriers and singularities?

Answer: By the time you finish this lesson you should have a good understanding of the procedure that links barriers to singularities. In this course, the theoretical justification of this basic procedure will be left a mystery in hopes of luring you to penetrate deeper into mathematics. You can continue your mathematical studies by taking a course called "Complex Variables" in which you will learn the theory behind the procedure. In the meantime, use the procedure with vigor and joy. Few calculus students not taking Calculus&*Mathematica* know about this stuff. Walk down the hallway and amaze some of the students in the traditional course.

■ B.2) Why you don't run into barriers when you approximate e^x, $\sin[x]$, and $\cos[x]$

B.2.a) Look at these attempts to approximate $f[x] = e^x$ by early parts of its expansion in powers of x:

```
In[45]:=
  Clear[f,approx,x]; f[x_] = E^x;
  approx[x_] = Normal[Series[f[x],{x,0,3}]];
  Plot[{f[x],approx[x]},{x,-4,4},
  PlotStyle->{{Thickness[0.02],Blue},
  {Thickness[0.01],Red}},PlotRange->
  {-10,E^4},AxesLabel->{"x",""},
  PlotLabel->"Powers of x centered at {0,0}"];
```

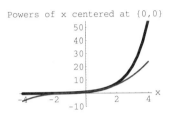

The approximation breaks down on the left and the right. Try to recover by using more of the expansion:

```
In[46]:=
  Clear[approx];
  approx[x_] = Normal[Series[f[x],{x,0,6}]];
  Plot[{f[x],approx[x]},{x,-4,4},
  PlotStyle->{{Thickness[0.02],Blue},
  {Thickness[0.01],Red}},
  PlotRange->{-10,E^4},
  AxesLabel->{"x",""},PlotLabel->
  "Powers of x centered at {0,0}"];
```

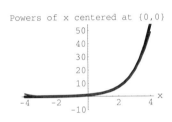

Again:

```
In[47]:=
  Clear[approx];
  approx[x_] = Normal[Series[f[x],{x,0,10}]];
  Plot[{f[x],approx[x]},{x,-4,4},
  PlotStyle->{{Thickness[0.02],Blue},
  {Thickness[0.01],Red}},PlotRange->{-10,E^4},
  AxesLabel->{"x",""},PlotLabel->
  "Powers of x centered at {0,0}"];
```

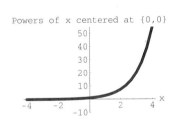

No signs of any barriers here.

> How do you explain why, when you go with $f[x] = e^x$, there are no barriers?

Answer: The key to learning why there are no barriers is to look for complex singularities of $f[x] = e^x$ and its derivatives. Fortunately, all of the higher derivatives of $f[x]$ are e^x.

```
In[48]:=
  Clear[x,f]; f[x_] = E^x;
  {f[x],f'[x],f''[x],f'''[x],f''''[x]}
```

Out[48]=
$\{E^x, E^x, E^x, E^x, E^x\}$

Looking for a barrier amounts to finding a complex singularity of e^z.

Taking $z = x + iy$, look at
$$e^z = e^{x+iy} = e^x \cos[y] + i\, e^x \sin[y].$$

There are no x's or y's that make $e^x \cos[y]$ blow up. And there are no x's or y's that make $e^x \sin[y]$ blow up. The upshot: e^z has no complex singularities. This is why you encounter no barriers when you try to approximate e^x by the early parts of its expansion in powers of $(x - b)$.

B.2.b.i) Look at these attempts to approximate $f[x] = \sin[x]$ by early parts of its expansion in powers of x:

In[49]:=
```
Clear[f,approx,x]; f[x_] = Sin[x];
approx[x_] = Normal[Series[f[x],{x,0,7}]];
Plot[{f[x],approx[x]},{x,-8,8},
PlotStyle->{{Thickness[0.02],Blue},
{Thickness[0.01],Red}},PlotRange->{-1.5,1.5},
AxesLabel->{"x",""},PlotLabel->
"Powers of x centered at {0,0}"];
```

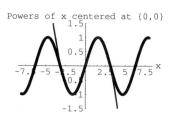

The approximation breaks down on the left and the right. Try to recover by using more of the expansion:

In[50]:=
```
Clear[approx];
approx[x_] = Normal[Series[f[x],{x,0,9}]];
Plot[{f[x],approx[x]},{x,-8,8},
PlotStyle->{{Thickness[0.02],Blue},
{Thickness[0.01],Red}},PlotRange->{-1.5,1.5},
AxesLabel->{"x",""},PlotLabel->
"Powers of x centered at {0,0}"];
```

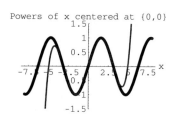

Again:

In[51]:=
```
Clear[approx];
approx[x_] = Normal[Series[f[x],{x,0,21}]];
Plot[{f[x],approx[x]},{x,-8,8},
PlotStyle->{{Thickness[0.02],Blue},
{Thickness[0.01],Red}},PlotRange->{-1.5,1.5},
AxesLabel->{"x",""},PlotLabel->
"Powers of x centered at {0,0}"];
```

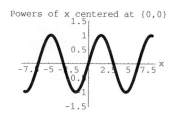

No signs of any barriers here.

Basics (B.2)

> How do you explain why, when you go with $f[x] = \sin[x]$, there are no barriers?

Answer: The key to learning why there are no barriers is to look for complex singularities of $f[x] = \sin[x]$ and its derivatives. All the higher derivatives of $f[x]$ are $\cos[x]$, $-\sin[x]$, $-\cos[x]$, and $\sin[x]$.

In[52]:=
```
Clear[x,f,k]; f[x_] = Sin[x]; Table[D[f[x],{x,k}],{k,1,12}]
```
Out[52]=
```
{Cos[x], -Sin[x], -Cos[x], Sin[x], Cos[x], -Sin[x],
  -Cos[x], Sin[x], Cos[x], -Sin[x], -Cos[x], Sin[x]}
```

Looking for a barrier amounts to finding complex singularitities of $\sin[z]$ and $\cos[z]$. To handle this, look at:

In[53]:=
```
ComplexExpand[I (E^(- I z) - E^(I z))/2]
```
Out[53]=
```
Sin[z]
```
and

In[54]:=
```
ComplexExpand[(E^(I z) + E^(-I z))/2]
```
Out[54]=
```
Cos[z]
```

This tells you that
$$\sin[z] = i \frac{e^{-iz} - e^{iz}}{2}$$
and
$$\cos[z] = \frac{e^{iz} + e^{-iz}}{2}.$$

Now you're in good shape, because you know that e^z has no complex singularities. This tells you that neither e^{iz} nor e^{-iz} has any complex singularities. Consequently neither $\sin[z] = i\left(e^{-iz} - e^{iz}\right)/2$ nor $\cos[z] = \left(e^{iz} + e^{-iz}\right)/2$ has any complex singularities. This is why you encounter no barriers when you try to approximate $\sin[x]$ by the early parts of its expansion.

B.2.b.ii) Look at these attempts to approximate $f[x] = \cos[x]$ by early parts of its expansion in powers of x:

In[55]:=
```
Clear[f,approx,x]; f[x_] = Cos[x];
approx[x_] = Normal[Series[f[x],{x,0,8}]];
Plot[{f[x],approx[x]},{x,-8,8},
PlotStyle->{{Thickness[0.02],Blue},
{Thickness[0.01],Red}},PlotRange->{-1.5,1.5},
AxesLabel->{"x",""},PlotLabel->
"Powers of x centered at {0,0}"];
```

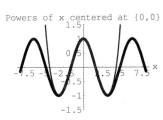

The approximation breaks down on the left and the right. Try to recover by using more of the expansion:

In[56]:=
```
Clear[approx];
approx[x_] = Normal[Series[f[x],{x,0,14}]];
Plot[{f[x],approx[x]},{x,-8,8},PlotStyle->
  {{Thickness[0.02],Blue},{Thickness[0.01],Red}},
  PlotRange->{-1.5,1.5},AxesLabel->{"x",""},
  PlotLabel->"Powers of centered at {0,0}"];
```

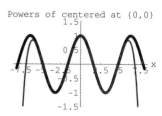

Again:

In[57]:=
```
Clear[approx];
approx[x_] = Normal[Series[f[x],{x,0,22}]];
Plot[{f[x],approx[x]},{x,-8,8},PlotStyle->
  {{Thickness[0.02],Blue},{Thickness[0.01],Red}},
  PlotRange->{-1.5,1.5},AxesLabel->{"x",""},
  PlotLabel->"Powers of x centered at {0,0}"];
```

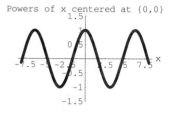

No signs of any barriers here.

> How do you explain why, when you go with $f[x] = \cos[x]$, there are no barriers?

Answer: The key to learning why there are no barriers is to look for complex singularities of $f[x] = \cos[x]$ and its derivatives. All the higher derivatives of $f[x]$ are $-\sin[x]$, $-\cos[x]$, $\sin[x]$, and $-\cos[x]$.

In[58]:=
```
Clear[x,f,k]; f[x_] = Cos[x];
Table[D[f[x],{x,k}],{k,1,12}]
```
Out[58]=
{-Sin[x], -Cos[x], Sin[x], Cos[x], -Sin[x], -Cos[x],
  Sin[x], Cos[x], -Sin[x], -Cos[x], Sin[x], Cos[x]}

Looking for a barrier amounts to finding complex singularities of $\sin[z]$ and $\cos[z]$. But in part B.2.b.i) above, you saw that neither $\sin[z]$ nor $\cos[z]$ has any complex singularities. This is the reason you run into no barriers when you try to approximate $\cos[x]$ by early terms of its expansions.

■ **B.3)** Why some functions like $x^{9/2}$ don't have expansions in powers of x

Here is the early part of the expansion of $f[x] = x^{9/2}$ in powers of $(x - b)$:

In[59]:=
```
Clear[f,x,b]; f[x_] = x^(9/2);
Normal[Series[f[x],{x,b,3}]]
```

Out[59]=

$$b^{9/2} + \frac{9 b^{7/2} (-b + x)}{2} + \frac{63 b^{5/2} (-b + x)^2}{8} + \frac{105 b^{3/2} (-b + x)^3}{16}$$

So far, so good. Now look at more of the expansion:

In[60]:=
```
Normal[Series[f[x],{x,b,6}]]
```
Out[60]=

$$b^{9/2} + \frac{9 b^{7/2} (-b + x)}{2} + \frac{63 b^{5/2} (-b + x)^2}{8} + \frac{105 b^{3/2} (-b + x)^3}{16} +$$
$$\frac{315 \text{Sqrt}[b] (-b + x)^4}{128} + \frac{63 (-b + x)^5}{256 \text{Sqrt}[b]} - \frac{21 (-b + x)^6}{1024 \, b^{3/2}}$$

Whoops. Those later terms have powers of b in the denominator.

When you take $b = 0$, you've got big problems, because you can't divide by 0. The upshot: $f[x] = x^{9/2}$ has no expansion in powers of x.

> Explain this sorry fact.

Answer: Again it's a matter of singularities. Look at $f[x]$ and some of its higher derivatives:

In[61]:=
```
Table[D[f[x],{x,k}],{k,0,9}]
```
Out[61]=

$$\left\{ x^{9/2},\ \frac{9 x^{7/2}}{2},\ \frac{63 x^{5/2}}{4},\ \frac{315 x^{3/2}}{8},\ \frac{945 \text{Sqrt}[x]}{16},\right.$$
$$\left.\frac{945}{32 \text{Sqrt}[x]},\ \frac{-945}{64 x^{3/2}},\ \frac{2835}{128 x^{5/2}},\ \frac{-14175}{256 x^{7/2}},\ \frac{99225}{512 x^{9/2}} \right\}$$

The first few derivatives do not blow up at $x = 0$, but the higher derivatives do blow up at $x = 0$. This tells you that when you expand in powers of x, the barriers pop up at $x = 0$, which is the center of your plotting interval. This is why $f[x] = x^{9/2}$ has no expansion in powers of x.

■ B.4) Barriers and convergence intervals

B.4.a.i) Fix a function $f[x]$ and a point b on the x-axis. Proceed as in B.1) above to find a number R such that the expansion of $f[x]$ in powers of $(x - b)$ has barriers at $b - R$ and $b + R$. Most all the good old math folks say that the expansion of $f[x]$ in

Lesson 4.05 Barriers to Convergence

powers of $(x - b)$ converges to $f[x]$ for x's with $b - R < x < b + R$. They call the interval $(b - R, b + R)$ the "convergence interval for the expansion of $f[x]$ in powers of $(x - b)$." Lots of Calculus&*Mathematica* students call the convergence interval $(b - R, b + R)$ the "cohabitation interval."

> Why do they say this?

Answer: Once you have located the barriers at $b - R$ and $b + R$, you know that using more and more of the expansion in powers of $(x - b)$ results in better and better high-quality approximations of $f[x]$ for x's with $b - R < x < b + R$.

B.4.a.ii) > What is the convergence interval for functions like e^x, $\sin[x]$, or $\cos[x]$ that have no barriers?

Answer: That's easy. The convergence interval for this kind of function is $(-\infty, \infty)$. This amounts to saying that the expansion of $f[x]$ in powers of $(x - b)$ converges to $f[x]$ for any old x you take.

B.4.b.i) The expansion of e^x in powers of x is

$$1 + x + \frac{x^2}{2!} + \frac{x^3}{3!} + \cdots + \frac{x^k}{k!} + \cdots$$

Put:

In[62]:=
```
Clear[expan,x,k,m]; expan[x_,m_] = Sum[x^k/k!,{k,0,m}];
expan[x,9]
```

Out[62]=

$$1 + x + \frac{x^2}{2} + \frac{x^3}{6} + \frac{x^4}{24} + \frac{x^5}{120} + \frac{x^6}{720} + \frac{x^7}{5040} + \frac{x^8}{40320} + \frac{x^9}{362880}$$

expan$[x, m]$ gives you the expansion of e^x through the mth degree term. Now look at these:

In[63]:=
```
{N[expan[1,3],20],N[E,20]}
```

Out[63]=
```
{2.6666666666666666667, 2.718281828459045235}
```

In[64]:=
```
{N[expan[1,5],20],N[E,20]}
```

Out[64]=
```
{2.7166666666666666667, 2.718281828459045235}
```

In[65]:=
```
{N[expan[1,10],20],N[E,20]}
```

Out[65]=
   {2.7182818011463844797, 2.718281828459045235}

In[66]:=
   {N[expan[1,20],20],N[E,20]}

Out[66]=
   {2.7182818284590452353, 2.718281828459045235}

In[67]:=
   {N[expan[1,40],20],N[E,20]}

Out[67]=
   {2.7182818284590452354, 2.718281828459045235}

> What's going on here?

Answer: You are watching the expansion of e^x in powers of x converging to e^x in the special case $x = 1$. The expansion of e^x converges to e^x on any interval, so when you take the expansion with $x = 1$ and run it out far enough, the result converges to e^1. Watch it happen to incredible accuracy:

In[68]:=
   {N[expan[1,50],70],N[E,70]}

Out[68]=
   {2.7182818284590452353602874713526624977572470936999595749669676277 23419,
   2.7182818284590452353602874713526624977572470936999595749669676277 24077}

Watch it happen for $x = 2$:

In[69]:=
   {N[expan[2,60],70],N[E^2,70]}

Out[69]=
   {7.3890560989306502272304274605750078131803155705518473240871278225 1788,
   7.3890560989306502272304274605750078131803155705518473240871278225 2257}

Watch it happen for $x = 3$:

In[70]:=
   {N[expan[3,60],70],N[E^3,70]}

Out[70]=
   {20.085536923187667740928529654581717896987907838554150144115656716 94949,
   20.085536923187667740928529654581717896987907838554150144378934229 69885}

Play with other x's.

B.4.b.ii) Fix a function $f[x]$ and a point b on the x-axis. Proceed as in B.1) above to find a number R such that the expansion of $f[x]$ in powers of $(x - b)$ has barriers at $b - R$ and $b + R$. All folks say that the infinite sum of all the terms of the expansion in powers of $(x - b)$ equals $f[x]$ for $b - R < x < b + R$.

> Why do they say this?

Answer: You know that using more and more of the expansion in powers of $(x - b)$ results in better and better approximations of $f[x]$ for x's with $b - R < x < b + R$. Consequently, when you use all the terms and add them up, you hit $f[x]$ right on the nose. For instance,

$$e^x = 1 + x + \frac{x^2}{2} + \frac{x^3}{3!} + \cdots + \frac{x^k}{k!} + \cdots \quad \text{for } -\infty < x < \infty.$$

And

$$\frac{1}{1-x} = 1 + x + x^2 + x^3 + x^4 + \cdots + x^k + \cdots$$

for x's with $-1 < x < 1$.

Tutorials

■ T.1) Convergence intervals

T.1.a) Determine the convergence interval for the expansion of

$$f[x] = \frac{e^{-x}}{(5 + 8x + 4x^2 + x^3)}$$

in powers of $(x - 2)$. Illustrate with a plot.

Answer: The key to determining where the barrier comes from depends on singularities and complex numbers. The function is $f[x] = e^{-x}/(5 + 8x + 4x^2 + x^3)$. No barriers come from the numerator because it has no complex singularities. Consequently, the only possible barriers can come from places where the denominator is 0.

In[1]:=
```
Clear[denominator,x,z]
denominator[x_] = 5 + 8 x + 4 x^2 + x^3;
singularities = Solve[denominator[z] == 0,z]
```

Out[1]=

$$\{\{z \to -1\}, \{z \to \frac{-3 - I\, \text{Sqrt}[11]}{2}\}, \{z \to \frac{-3 + I\, \text{Sqrt}[11]}{2}\}\}$$

The complex singularities are -1, $-3/2 + i\sqrt{1}\,1/2$ and, $-3/2 - i\sqrt{1}/2$. You are expanding in powers of $(x - 2)$, so the center of the convergence interval is $\{2, 0\}$. Plot

$$\{-1, 0\}, \left\{\frac{-3}{2}, \frac{\sqrt{11}}{2}\right\} \quad \text{and} \quad \left\{\frac{-3}{2}, \frac{-\sqrt{11}}{2}\right\}$$

together with the center of the convergence interval $\{2, 0\}$:

In[2]:=
```
sing1 = {-3/2,Sqrt[11]/2};
sing2 = {-3/2,-Sqrt[11]/2};
sing3 = {-1,0}; center = {2,0};
points = {Graphics[{Red,PointSize[0.04],
Point[sing1]}],Graphics[{Red,PointSize[0.04],
Point[sing2]}],Graphics[{Red,PointSize[0.04],
Point[sing3]}],Graphics[{Blue,PointSize[0.04],
Point[center]}]}; h=0.3;
labels = {Graphics[Text["sing1",sing1-{0,h}]],
Graphics[Text["sing2",sing2+{0,h}]],
Graphics[Text["sing3",sing3+{0,h}]],
Graphics[Text["center",center+{0,h}]]};
Show[points,labels,Axes->True,PlotRange->All];
```

Run line segments from the center of the plotting interval to each singularity:

In[3]:=
```
lines = {Graphics[{Thickness[0.01],
Line[{center,sing1}]}],Graphics[{Thickness[0.01],
Line[{center,sing2}]}],Graphics[{Thickness[0.01],
Line[{center,sing3}]}]};
Show[lines,points,labels,
Axes->True,PlotRange->All];
```

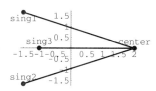

The top line connects the points:

In[4]:=
```
{center,sing1}
```

Out[4]=
$$\{\{2, 0\}, \{-(\frac{3}{2}), \frac{\text{Sqrt}[11]}{2}\}\}$$

Its length is:

In[5]:=
```
length1 = N[Sqrt[(-3/2 - 2)^2 + (Sqrt[11]/2 - 0)^2]]
```

Out[5]=
3.87298

The bottom line connects the points:

In[6]:=
```
{center,sing2}
```

Out[6]=
$$\{\{2, 0\}, \{-(\frac{3}{2}), \frac{-\text{Sqrt}[11]}{2}\}\}$$

Its length is:

In[7]:=
```
length2 = N[Sqrt[(-3/2 - 2)^2 + (-Sqrt[11]/2 - 0)^2]]
```

Out[7]=
3.87298

The middle line connects the points:

In[8]:=
{center,sing3}

Out[8]=
{{2, 0}, {-1, 0}}

Its length is:

In[9]:=
length3 = N[Sqrt[(-1 - 2)^2 + (0 - 0)^2]]

Out[9]=
3.

Compare the lengths:

In[10]:=
{length1,length2,length3}

Out[10]=
{3.87298, 3.87298, 3.}

For this problem, $R = 3$ is the magic number. The expansion of $f[x]$ in powers of $(x - 2)$ converges to $f[x]$ for $2 - 3 < x < 2 + 3$. The barriers are at $x = -1$ and $x = 5$. See the action on the convergence interval:

In[11]:=
```
Clear[f,approx,x]; f[x_] = (E^(-x))/(5 + 8 x + 4 x^2 + x^3);
b = 2; R = 3; leftbarrier = b - R; rightbarrier = b + R;
approx[x_] = Normal[Series[f[x],{x,b,18}]];
seriousplot = Plot[{f[x],approx[x]},{x, b - R ,b + R},
PlotStyle->{{Thickness[0.02],Blue},{Thickness[0.01],Red}},PlotRange->{-0.3,4},
Epilog->{{Blue,PointSize[0.04],Point[{b,0}]},Text["expansion point",{b,0},{0,-2}]},
DisplayFunction->Identity]; barriers = {Graphics[{Line[{{leftbarrier,-0.3},
{leftbarrier,4}}]}],Graphics[{Line[{{rightbarrier,-0.3},{rightbarrier,4}}]}]};
```

In[12]:=
```
Show[seriousplot,barriers,
AxesLabel->{"x",""},PlotLabel->
"Action on the convergence interval",
DisplayFunction->$DisplayFunction];
```

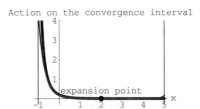

T.1.b) Determine the convergence interval for the expansion of

$$f[x] = \frac{\sin[x]}{\cos[x]}$$

in powers of x. Illustrate with a plot.

Answer: The key to determining where the barrier comes from depends on singularities and complex numbers. The function is $f[x] = \sin[x]/\cos[x]$. No barriers come from the numerator because $\sin[z]$ has no complex singularities. Consequently, the only possible barriers come from places where the denominator is 0.

In[13]:=
```
Clear[denominator,x,z]
denominator[x_] = Cos[x];
singularities = Solve[denominator[z] == 0,z]
```
Solve::ifun:
   Warning: Inverse functions are being used by Solve, so
     some solutions may not be found.

Out[13]=
$$\{\{z \to \frac{Pi}{2}\}\}$$

The complex singularities are $\pi/2, -\pi/2, 3\pi/2, -3\pi/2, \ldots$. You are expanding in powers of x, so the center of the convergence interval is $\{0,0\}$. The singularities closest to $\{0,0\}$ are $\pi/2$ and $-\pi/2$. For this problem, the magic number is $R = \pi/2$. The expansion of $f[x]$ in powers of x converges to $f[x]$ for $0 - \pi/2 < x < 0 + \pi/2$. The barriers are at $x = -\pi/2$ and $x = \pi/2$. See the action on the convergence interval:

In[14]:=
```
Clear[f,approx,x]; f[x_] = Sin[x]/Cos[x]; b = 0; R = Pi/2;
leftbarrier = b - R; rightbarrier = b + R;
approx[x_] = Normal[Series[f[x],{x,b,9}]];
seriousplot = Plot[{f[x],approx[x]},{x, b - R ,b + R},PlotStyle->
{{Thickness[0.02],Blue},{Thickness[0.01],Red}},DisplayFunction->Identity];
barriers = {Graphics[{Line[{{leftbarrier,-20},{leftbarrier,20}}]}],
Graphics[{Line[{{rightbarrier,-20},{rightbarrier,20}}]}]};
```

In[15]:=
```
Show[seriousplot,barriers,
AxesLabel->{"x",""},
PlotLabel->
"Action on the convergence interval",
DisplayFunction->$DisplayFunction];
```

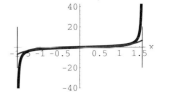

More of the action:

In[16]:=
```
Clear[approx]; approx[x_] = Normal[Series[f[x],{x,b,21}]];
moreseriousplot = Plot[{f[x],approx[x]},{x, b - R ,b + R},
PlotStyle->{{Thickness[0.02],Blue},{Thickness[0.01],Red}},
DisplayFunction->Identity];
```

Lesson 4.05 Barriers to Convergence

In[17]:=
```
Show[moreseriousplot,barriers,
AxesLabel->{"x",""},
PlotLabel->
"Action on the convergence interval",
DisplayFunction->$DisplayFunction];
```

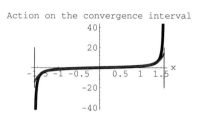

T.1.c) Determine the convergence interval for the expansion of
$$f[x] = e^x \sin[x]$$
in powers of $(x - b)$.

Answer: The expansion of $e^x \sin[x]$ in powers of $(x - b)$ is the expansion of e^{-x} in powers of $(x - b)$ times the expansion of $\sin[x]$ in powers of $(x - b)$.

The expansion of e^x in powers of $(x - b)$ converges to e^x for any x. The expansion of $\sin[x]$ in powers of $(x - b)$ converges to $\sin[x]$ for any x. So, the magic number is $R = \infty$. Consequently, the expansion of $e^x \sin[x]$ in powers of $(x - b)$ converges to $e^x \sin[x]$ for any x.

T.1.d) Determine the convergence interval for the expansion of
$$f[x] = \sqrt{2 + x}$$
in powers of $(x - 3)$.

Answer: The key to determining where the barrier comes from depends on singularities and complex numbers. The function is $f[x] = \sqrt{2 + x}$. $f[x]$ has no blatant blow-ups (singularities), but $f'[x]$ does:

In[18]:=
```
Clear[f,x]; f[x_] = Sqrt[2 + x]; f'[x]
```
Out[18]=
$$\frac{1}{2 \, \text{Sqrt}[2 + x]}$$

$f'[x]$ blows sky high at $x = -2$:

In[19]:=
```
f'[-2]
```

Power::infy: Infinite expression $\dfrac{1}{\text{Sqrt}[0]}$ encountered.

Out[19]=
ComplexInfinity

This sets a barrier at $x = -2$. And because you are expanding in powers of $(x-3)$, the center of the convergence interval is $\{3, 0\}$. This tells you that the magic number is $R = 5$.

The upshot: The convergence interval for the expansion of $f[x] = \sqrt{2+x}$ in powers of $(x-3)$ is $3 - 5 < x < 3 + 5$. This is the same as $-2 < x < 8$.

See it happen:

In[20]:=
```
b = 3; R = 5; leftbarrier = b - R; rightbarrier = b + R;
approx[x_] = Normal[Series[f[x],{x,b,12}]];
seriousplot = Plot[{f[x],approx[x]},{x, b - R - 1 ,b + R + 1},
PlotStyle->{{Thickness[0.02],Blue},{Thickness[0.01],Red}},
PlotRange->{-0.3,4},Epilog->{{Blue,PointSize[0.04],Point[{b,0}]},
Text["expansion point",{b,0},{0,-2}]},DisplayFunction->Identity];
barriers = {Graphics[{Red,Line[{{leftbarrier,-0.3},{leftbarrier,4}}]}],
Graphics[{Red,Line[{{rightbarrier,-0.3},{rightbarrier,4}}]}]};
```

In[21]:=
```
Show[seriousplot,barriers,
AxesLabel->{"x",""},
PlotLabel->
"Action on the convergence interval",
DisplayFunction->$DisplayFunction];
```

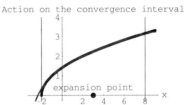

■ T.2) The convergence intervals for $f[x]$, $f'[x]$ and $\int_a^x f[t]\,dt$ are all the same

T.2.a) Given a function $f[x]$ and a number a, try to explain how you know that the convergence intervals for $f[x], f'[x]$ and $\int_a^x f[t]\,dt$ are all the same?

Answer: To find the convergence intervals for $f[x]$, you look for complex singularities of $f[x]$ and its higher derivatives: $f'[x], f''[x], f'''[x], \ldots$. The fact is that once one of these high derivatives hits a singularity, then all the even higher derivatives hit the same singularity. To find the convergence intervals for $f'[x]$, you look for complex singularities of $f'[x]$ and its higher derivatives: $f''[x], f'''[x], f''''[x], \ldots$ In view of what was said above, the same singularities will emerge from this list as emerge from the list $\{f[x], f'[x], f''[x], f'''[x], \ldots\}$ The upshot: The convergence intervals for $f[x]$ and for $f'[x]$ are the same.

To find the convergence intervals for
$$g[x] = \int_a^x f[t]\,dt,$$
you look for complex singularities of $g[x]$ and its derivatives:
$$\{g[x], g'[x], g''[x], g'''[x], g''''[x], \ldots\} = \{g[x], f[x], f'[x], f''[x], f'''[x], \ldots\}.$$

In view of what was said above, the same singularities will emerge from this list as emerge from the list $\{f[x], f'[x], f''[x], f'''[x], \ldots\}$. The upshot: The convergence intervals for $f[x]$ and for $g[x] = \int_a^x f[t]\, dt$ are the same.

T.2.b) Look at this:

In[22]:=
```
Clear[x]; Together[D[1/(1 + x),x]]
```

Out[22]=

$-(1 + x)^{-2}$

This tells you that the derivative of $1/(1-x)$ is $-1/(1-x)^2$.

> What does this tell you about the convergence intervals of $-1/(1-x)^2$?

Answer: It tells you that the convergence intervals of $-1/(1-x)^2$ are the same as the convergence intervals of $1/(1-x)$. For example, the convergence interval for the expansion of $-1/(1-x)^2$ in powers of $(x-4)$ is the interval $1 < x < 7$.

T.2.c) Look at this:

In[23]:=
```
Clear[x,t]; Integrate[1/(1 + t^2),{t,0,x}]
```

Out[23]=
```
ArcTan[x]
```

This tells you that

$$\int_0^x \frac{1}{1+t^2}\, dt = \arctan[x].$$

> What does this tell you about the convergence intervals of $\arctan[x]$?

Answer: It tells you that the convergence intervals of $\arctan[x]$ are the same as the convergence intervals of $1/(1+x^2)$. For example, the convergence interval for the expansion of $\arctan[x]$ powers of x is the interval $-1 < x < 1$.

■ **T.3)** $1/(1-x) = 1 + x + x^2 + \cdots + x^k + \cdots$ for $-1 < x < 1$

T.3.a) Explain how you know that

$$\frac{1}{1-x} = 1 + x + x^2 + x^3 + \cdots + x^k + \cdots \qquad \text{for } -1 < x < 1.$$

Answer: The function is $f[x] = 1/(1-x)$. The only possible barriers come from places where the denominator is 0, and this happens at $x = 1$. You are expanding in powers of x, so the center of the convergence interval is $\{0, 0\}$. The lone singularity is at $x = 1$. For this problem, the magic number is $R = 1$. This means that the convergence interval of the expansion of $1/(1-x)$ in powers of x is $0 - 1 < x < 0 + 1$. In other words,

$$\frac{1}{1-x} = 1 + x + x^2 + x^3 + \cdots + x^k + \cdots \qquad \text{for } -1 < x < 1.$$

See it happen:

In[24]:=
```
Clear[f,approx,x]; f[x_] = 1/(1 - x);
b = 0; R = 1; leftbarrier = b - R; rightbarrier = b + R;
approx[x_] = Normal[Series[f[x],{x,b,20}]];
seriousplot = Plot[{f[x],approx[x]},{x, b - R ,b + R},
PlotStyle->{{Thickness[0.02],Blue},{Thickness[0.01],Red}},
DisplayFunction->Identity];
barriers = {Graphics[{Line[{{leftbarrier,-0.5},{leftbarrier,10}}]}],
Graphics[{Line[{{rightbarrier,-0.5},{rightbarrier,10}}]}]};
```

In[25]:=
```
Show[seriousplot,barriers,
AxesLabel->{"x",""},PlotRange->{-0.5,10},
PlotLabel->
"Action on the convergence interval",
DisplayFunction->$DisplayFunction];
```

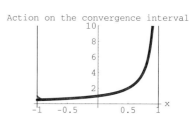

T.3.b) Use what you know about the convergence interval of the expansion of $1/(1-x)$ in powers of x to determine the convergence interval of the expansion of

$$\frac{1}{1+2x^6}$$

in powers of x.

Answer:

$$\frac{1}{1-x} = 1 + x + x^2 + x^3 + \cdots + x^k + \cdots \qquad \text{for } -1 < x < 1.$$

Change x to $-2x^6$ throughout to learn that

$$\frac{1}{1+2x^6} = 1 - 2x^6 + 2^2 x^{12} + \cdots + (-1)^k 2^k x^{6k} + \cdots \qquad \text{for } -1 < -2x^6 < 1.$$

Saying that

$$-1 < -2x^6 < 1$$

is the same as saying
$$\frac{1}{2} > x^6 > -\frac{1}{2}.$$

The upshot: The convergence interval of the expansion of $1/(1+2x^6)$ in powers of x is
$$-\left(\frac{1}{2}\right)^{1/6} < x < \left(\frac{1}{2}\right)^{1/6}.$$

Watch it happen:

In[26]:=
```
Clear[f,approx,x]; f[x_] = 1/(1 + 2 x^6); b = 0;
R = (1/2)^(1/6); leftbarrier = b - R; rightbarrier = b + R;
approx[x_] = Normal[Series[f[x],{x,b,20}]];
seriousplot = Plot[{f[x],approx[x]},{x, b - R ,b + R},
 PlotStyle->{{Thickness[0.02],Blue},{Thickness[0.01],Red}},DisplayFunction->Identity];
barriers = {Graphics[{Line[{{leftbarrier,-0.2},{leftbarrier,1.2}}]}],
 Graphics[{Line[{{rightbarrier,-0.2},{rightbarrier,1.2}}]}]};
```

In[27]:=
```
Show[seriousplot,barriers,
AxesLabel->{"x",""},PlotRange->{-0.2,1.2},
PlotLabel->
"Action on the convergence interval",
DisplayFunction->$DisplayFunction];
```

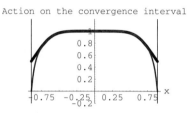

Sweet.

T.3.c) This problem appears only in the electronic version.

■ T.4) Infinite sums of numbers

T.4.a) Find the infinite sum and illustrate the convergence of
$$1 + \left(\frac{1}{2}\right) + \left(\frac{1}{2}\right)^2 + \left(\frac{1}{2}\right)^3 + \left(\frac{1}{2}\right)^4 + \cdots + \left(\frac{1}{2}\right)^k + \cdots$$

Answer: Recall:
$$\frac{1}{1-x} = 1 + x + x^2 + \cdots + x^k + \cdots$$
provided $-1 < x < 1$. Plug in $x = 1/2$ to see that
$$1 + \left(\frac{1}{2}\right) + \left(\frac{1}{2}\right)^2 + \left(\frac{1}{2}\right)^3 + \left(\frac{1}{2}\right)^4 + \cdots + \left(\frac{1}{2}\right)^k + \cdots$$

is given by:

In[28]:=
    Clear[x]; 1/(1 - x)/.x->1/2

Out[28]=
    2

So, $2 = 1 + \left(\frac{1}{2}\right) + \left(\frac{1}{2}\right)^2 + \left(\frac{1}{2}\right)^3 + \cdots + \left(\frac{1}{2}\right)^k + \cdots$

Neat.

Watch it converge:

In[29]:=
    Clear[k]; N[Sum[(1/2)^k,{k,0,10}],12]

Out[29]=
    1.9990234375

In[30]:=
    N[Sum[(1/2)^k,{k,0,20}],12]

Out[30]=
    1.99999904633

In[31]:=
    N[Sum[(1/2)^k,{k,0,30}],12]

Out[31]=
    1.99999999907

In[32]:=
    N[Sum[(1/2)^k,{k,0,100}],40]

Out[32]=
    1.999999999999999999999999999999211139095

Zingo.

T.4.b) Find the infinite sum, and illustrate the convergence of

$$\left(\frac{1}{5}\right) - \left(\frac{1}{5}\right)^2 + \left(\frac{1}{5}\right)^3 - \left(\frac{1}{5}\right)^4 + \cdots + (-1)^k \left(\frac{1}{5}\right)^{k+1} + \cdots$$

Answer: Recall:
$$\frac{1}{1-x} = 1 + x + x^2 + x^3 + \cdots + x^k + \cdots$$
provided $-1 < x < 1$. So,
$$\frac{-x}{1-x} = -x - x^2 - x^3 - \cdots - x^{k+1} - \cdots \qquad \text{provided } -1 < x < 1.$$
Replacing x by $-x$ gives
$$\frac{x}{1+x} = x - x^2 + x^3 + \cdots + (-1)^k x^{k+1} + \cdots \qquad \text{provided } -1 < x < 1.$$

Plug in $x = 1/5$ to see that

$$\left(\frac{1}{5}\right) - \left(\frac{1}{5}\right)^2 + \left(\frac{1}{5}\right)^3 - \left(\frac{1}{5}\right)^4 + \cdots + (-1)^k \left(\frac{1}{5}\right)^{k+1} + \cdots$$

is given by:

In[33]:=
```
Clear[x]; x/(1 + x)/.x->1/5
```
Out[33]=
$$\frac{1}{6}$$

So,

$$\frac{1}{6} = \left(\frac{1}{5}\right) - \left(\frac{1}{5}\right)^2 + \left(\frac{1}{5}\right)^3 + \cdots + (-1)^k \left(\frac{1}{5}\right)^{k+1} + \cdots$$

Watch it converge:

In[34]:=
```
Clear[k]; N[{Sum[(-1)^(k+2) (1/5)^(k+1),{k,0,5}],1/6},20]
```
Out[34]=
{0.166656, 0.16666666666666666667}

In[35]:=
```
N[{Sum[(-1)^(k+2) (1/5)^(k+1),{k,0,10}],1/6},20]
```
Out[35]=
{0.16666667008, 0.16666666666666666667}

In[36]:=
```
N[{Sum[(-1)^(k+2) (1/5)^(k+1),{k,0,20}],1/6},20]
```
Out[36]=
{0.16666666666666701619, 0.16666666666666666667}

In[37]:=
```
N[{Sum[(-1)^(k+2) (1/5)^(k+1),{k,0,30}],1/6},30]
```
Out[37]=
{0.166666666666666666666702458061, 0.166666666666666666666666666667}

In[38]:=
```
N[{Sum[(-1)^(k+2) (1/5)^(k+1),{k,0,40}],1/6},40]
```
Out[38]=
{0.1666666666666666666666666666666703317054259,
 0.1666666666666666666666666666666666666667}

That's furious convergence!

T.4.c) Find the infinite sum, and illustrate the convergence of

$$1 - \frac{1}{2!} + \frac{1}{4!} - \frac{1}{6!} + \frac{1}{8!} + \cdots + \frac{(-1)^k}{(2k)!} + \cdots$$

Answer:

$$\cos[x] = 1 - \frac{x^2}{2!} + \frac{x^4}{4!} - \frac{x^6}{6!} + \frac{x^8}{8!} + \cdots + \frac{(-1)^k x^k}{(2k)!} + \cdots$$

for any old x. Plug in $x = 1$ to see that

$$\cos[1] = 1 - \frac{1}{2!} + \frac{1}{4!} - \frac{1}{6!} + \frac{1}{8!} + \cdots + \frac{(-1)^k}{(2\,k)!} + \cdots$$

Watch it converge:

In[39]:=
```
Clear[k]; N[{Sum[((-1)^k)/(2 k)!,{k,0,5}],Cos[1]},12]
```
Out[39]=
```
{0.540302303792, 0.540302305868}
```

In[40]:=
```
N[{Sum[((-1)^k)/(2 k)!,{k,0,8}],Cos[1]},20]
```
Out[40]=
```
{0.54030230586813987732, 0.54030230586813971740}
```

In[41]:=
```
N[{Sum[((-1)^k)/(2 k)!,{k,0,12}],Cos[1]},20]
```
Out[41]=
```
{0.54030230586813971740, 0.54030230586813971740}
```

Math happens again!

■ T.5) Using the expansion of $1/(1-x)$ for drug dosing

T.5.a.i) A dose of a certain prescription drug into a human being results in an immediate infusion of G grams of the drug into the body fluids. It is known that t hours after one dose, the concentration in grams per liter of the drug resulting from that single dose is $(G/B)\,e^{-at}$, where a is a positive parameter that depends on the individual drug, and not on the dose. Here B measures the total body fluids in liters. Suppose the drug is taken regularly every p hours.

> Give a formula for the concentration of the drug in the body fluids immediately before the eighth dose.

Answer: The concentration in the body fluids immediately before the eighth dose is

$$\frac{G\,e^{-ap} + G\,e^{-2ap} + G\,e^{-3ap} + G\,e^{-4ap} + G\,e^{-5ap} + G\,e^{-6ap} + G\,e^{-7ap}}{B}.$$

T.5.a.ii) Assume this drug is taken every p hours on a regular basis for a very, very long time.

> What is the approximate concentration of the drug before each new dose? How is this information used in setting doses of drugs?

Answer: The approximate concentration of the drug before each new dose is the infinite sum,

$$\frac{Ge^{-ap} + Ge^{-2ap} + Ge^{-3ap} + Ge^{-4ap} + \cdots + Ge^{-kap} + \cdots}{B}$$

$$= \frac{G}{B}\left(e^{-ap} + e^{-2ap} + e^{-3ap} + e^{-4ap} + \cdots + e^{-kap} + \cdots\right).$$

This is the same as

$$\frac{G}{B}\left(x + x^2 + x^3 + x^4 + \cdots + x^k + \cdots\right) \quad \text{with } x = e^{-ap}.$$

Because a and p are positive, you are guaranteed that $0 < e^{-ap} < e^{-0} = 1$.

Recall that:

$$\frac{1}{1-x} = 1 + x + x^2 + x^3 + x^4 + x^5 + \cdots \quad \text{for } -1 < x < 1.$$

So

$$\frac{x}{1-x} = x + x^2 + x^3 + x^4 + x^5 + \cdots \quad \text{for } -1 < x < 1.$$

This tells you that approximate concentration of the drug before each new dose,

$$\frac{G}{B}\left(e^{-ap} + e^{-2ap} + e^{-3ap} + e^{-4ap} + e^{-5ap} + \cdots + e^{-kap} + \cdots\right)$$

is unmasked to be

$$\frac{G}{B}\frac{x}{1-x} \quad \text{with } x = e^{-ap}.$$

This is given by:

In[42]:=
```
Clear[a,p,x,G,B]; Simplify[(G/B) x/(1 - x)/.x->E^(-a p)]
```

Out[42]=
$$\frac{G}{-B + B E^{a\,p}}$$

In practice the pharmacologist knows the given number a, the maximum safe concentration S of the drug in the body fluids, and the number B of liters of body fluids in the recipient. With these data, the dose size G and the dose spacing p are set so that $G/(-B + Be^{ap}) < S$. This ensures that the concentration never exceeds the maximum safe concentration, S.

T.5.a.iii) Here is some information about the body fluid supply of college age males and females:

→ A college-age male in good shape, weighing K kilograms has about $0.68\,K$ liters of fluid in his body. Males in poor shape have less.

→ A college-age female in good shape, weighing K kilograms has about $0.65\,K$ liters of fluid in her body. Females in poor shape have less.

Now get down to brass tacks: A drug is to be prescribed in doses of 0.5 grams to a college-age female weighing 110 pounds. She is in good physical condition. One dose of this drug results in a body fluid concentration of $(0.5/B)\,e^{-0.12t}$ grams per liter t hours after the dose. The student is to take this drug every p hours, where p is set so that the concentration of this drug before each new dose never exceeds 0.011 grams per liter of her body fluids.

> Your job is to set the smallest safe value of p.

Answer: She weighs 110 pounds. In kilograms, this measures out to:

In[43]:=
```
Convert[110 PoundWeight,KilogramWeight]
```
Out[43]=
```
Convert[110 PoundWeight, KilogramWeight]
```

In liters, her body fluids measure out to:

In[44]:=
```
B = 0.65  49.8951
```
Out[44]=
```
32.4318
```

In this situation:

In[45]:=
```
G = 0.5; a = 0.12;
```

Over the long haul, with a spacing of p hours, the concentration of this drug in her blood immediately before a new dose will measure out to $G/(-B + Be^{ap})$ grams per liter:

In[46]:=
```
Clear[concentration,p]; concentration[p_] = G/(-B + B E^(a p))
```
Out[46]=
$$\frac{0.5}{-32.4318 + 32.4318\, E^{0.12\,p}}$$

Remember that p is to be set so that

$$\text{concentration}[p] \leq 0.011.$$

Here's a look:

```
In[47]:=
  maxsafe = 0.011;
  Plot[{concentration[p],maxsafe},{p,1,12},
  PlotStyle->{{Blue,Thickness[0.01]},{Red}},
  AxesLabel->{"p",
  "concentration before new dose"}];
```

She can take this drug every $p = 8$ hours. You can get a more precise estimate of the smallest safe p by looking at:

```
In[48]:=
  Solve[concentration[p] == maxsafe,p]
Solve::ifun:
  Warning: Inverse functions are being used by Solve, so
    some solutions may not be found.

Out[48]=
  {{p -> 7.30092}}
```

The smallest safe dosage interval is actually $p = 7.3$ hours, but no one can remember to take a drug every 7.3 hours, so the eyeball answer of $p = 8$ is the sensible answer here.

Give It a Try

Experience with the starred ($\star$) problems will be especially beneficial for understanding later lessons.

■ G.1) Convergence intervals*

G.1.a) Replace the question mark with the right response. The first two are done for you.

Function: $\longrightarrow$ Interval of convergence of expansion in powers of x:

$\cos[x^2] \longrightarrow (-\infty, \infty)$

$\dfrac{1}{1 + 3x^2} \longrightarrow \left(\dfrac{-1}{\sqrt{3}}, \dfrac{1}{\sqrt{3}}\right)$

$e^{-x^2} \longrightarrow (-?, ?)$

$\dfrac{1}{1 - x} \longrightarrow (-?, ?)$

$$\frac{1}{1+x} \longrightarrow (-?,?)$$

$$\frac{1}{1+4x} \longrightarrow (-?,?)$$

$$\frac{1}{1+4x^2} \longrightarrow (-?,?)$$

$$\arctan[2x] \longrightarrow (-?,?)$$

$$\arctan\left[\frac{x}{2}\right] \longrightarrow (-?,?)$$

$$\log[1+4x^2] \longrightarrow (-?,?)$$

$$\sin\left[x+\frac{\pi}{4}\right] \longrightarrow (-?,?)$$

$$\sqrt{1+\frac{x}{2}} \longrightarrow (-?,?)$$

$$e^{-x}\sin[x^2] \longrightarrow (-?,?)$$

$$\frac{\cos[x]}{1+x^4} \longrightarrow (-?,?)$$

■ G.2) Convergence intervals and plots★

Give the intervals of convergence of the expansion in powers of x for each of the following functions, then illustrate your results with plots.

G.2.a) $\arctan[x]$

G.2.b) $\dfrac{e^x}{5+x^2}$

G.2.c) $\dfrac{4x+16}{x^3-6x^2+10x-8}$

■ G.3) Sharing ink★

G.3.a) In each part, plot the function and enough of its expansion in powers of x on the same axes so that the two plots share the same ink on the indicated interval.

G.3.a.i) $\cos[x^2]$ on $[-2, 2]$

G.3.a.ii) e^{-x^2} on $[-3, 3]$

G.3.a.iii) $\sin[x]^2$ on $[-2, 2]$

G.3.b) Why is it impossible to do the same thing for $1/(1 + 2x^2)$ on $[-0.9, 0.9]$?

■ G.4) Infinite sums of numbers

In each of the following, when you are asked to illustrate the convergence, do it in the style used in the Tutorials.

G.4.a) Find the infinite sum and illustrate the convergence of

$$\left(\frac{1}{4}\right) + \left(\frac{1}{4}\right)^2 + \left(\frac{1}{4}\right)^3 + \left(\frac{1}{4}\right)^4 + \cdots + \left(\frac{1}{4}\right)^k + \cdots$$

G.4.b) Find the infinite sum and illustrate the convergence of

$$1 + \frac{1}{2} + \frac{1}{2!\,2^2} + \frac{1}{3!\,2^3} + \cdots + \frac{1}{k!\,2^k} + \cdots$$

G.4.c) Find the infinite sum and illustrate the convergence of

$$1 - \frac{1}{2!\,2} + \frac{1}{4!\,2^2} - \frac{1}{6!\,2^3} + \cdots + \frac{(-1)^n}{(2n)!\,2^n} + \cdots$$

G.4.d) Find the infinite sum and illustrate the convergence of

$$\frac{1}{2} - \left(\frac{1}{2}\right)^2 + \left(\frac{1}{2}\right)^3 - \left(\frac{1}{2}\right)^4 + \cdots + (-1)^{n+1}\left(\frac{1}{2}\right)^n + \cdots$$

G.4.e) Find the infinite sum and illustrate the convergence of
$$\frac{\pi}{2} - \frac{\pi^3}{2^3\, 3!} + \frac{\pi^5}{2^5\, 5!} + \cdots + \frac{(-1)^k\, \pi^{2k+1}}{2^{2k+1}\, (2k+1)!} + \cdots$$

G.4.f.i) Find the infinite sum and illustrate the convergence of
$$\frac{1}{3} + \frac{2}{3^2} + \frac{3}{3^3} + \frac{4}{3^4} + \cdots + \frac{k}{3^k} + \cdots$$

G.4.f.ii) Find the infinite sum and illustrate the convergence of
$$1 + \frac{2}{3^2} + \frac{3}{3^3} + \frac{4}{3^4} + \cdots + \frac{k}{3^k} + \cdots$$

■ G.5) Barriers resulting from splines

To this point, the focus has been on functions that are not splines. But when you work with a function that comes from splining two other functions, some unforeseen things can happen.

Here is what you get when you set constants a, b, and c so that when you knot at $\{1, e\}$, the spline $h[x]$ given by

$$\text{spline}[x] = f[x] = e^x \qquad \text{for } x \leq 1$$

and

$$\text{spline}[x] = g[x] = a + b\, \cos[\pi\, x] + c\, \sin[\pi\, x] \qquad \text{for } x > 1$$

has as much smoothness at the knot as it can:

In[1]:=
```
Clear[f,prelimg,g,x,a,b,c]; f[x_] = E^x;
prelimg[x_] = a + b Cos[Pi x] + c Sin[Pi x]
```
Out[1]=
```
a + b Cos[Pi x] + c Sin[Pi x]
```

In[2]:=
```
eqn1 = f[1] == prelimg[1]
```
Out[2]=
```
E == a - b
```

In[3]:=
```
eqn2 = f'[1] == prelimg'[1]
```
Out[3]=
```
E == -(c Pi)
```

Lesson 4.05 Barriers to Convergence

In[4]:=
```
eqn3 = f''[1] == prelimg''[1]
```
Out[4]=
$$E == b\, Pi^2$$

In[5]:=
```
solutions = Solve[{eqn1,eqn2,eqn3},{a,b,c}]
```
Out[5]=
$$\{\{a \to E + \frac{E}{Pi^2},\ b \to \frac{E}{Pi^2},\ c \to -(\frac{E}{Pi})\}\}$$

In[6]:=
```
g[x_] = prelimg[x]/.solutions[[1]]
```
Out[6]=
$$E + \frac{E}{Pi^2} + \frac{E\,Cos[Pi\,x]}{Pi^2} - \frac{E\,Sin[Pi\,x]}{Pi}$$

Here comes the plot of the spline:

In[7]:=
```
Clear[spline]
spline[x_] := f[x]/;x <= 1 ;
spline[x_] := g[x]/;x > 1;
splineplot = Plot[spline[x],{x,-3,3},
  PlotStyle->{{Thickness[0.02],Blue}},
  AxesLabel->{"x","spline[x]"},
  Epilog->{{Red,PointSize[0.04],Point[{1,E}]},
  Text["knot",{1,E},{1,-2}]}];
```

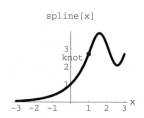

A nice, smooth spline. This function spline[x] is $f[x] = e^x$ to the left of the knot at $\{1, e\}$. Taylor's formula tells you that expanding spline[x] in powers of x is the same as expanding $f[x]$ in powers of x.

See what happens when you try for a serious approximation using the expansion of spline[x] in powers of x:

In[8]:=
```
b = 0; approx[x_] = Normal[Series[f[x],{x,b,12}]];
approxplot = Plot[approx[x],{x,b - 2.5,b + 2.5},
  PlotStyle->{{Thickness[0.01],Red}},
  DisplayFunction->Identity];
Show[splineplot,approxplot,
  DisplayFunction->$DisplayFunction];
```

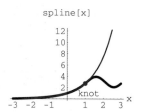

G.5.a.i) Experiment by using more of the expansion in powers of x. Do you think there's a barrier anywhere? If so, where?

> What do you think is the interval of convergence of the expansion of spline[x] in powers of x?

G.5.a.ii) See what happens when you try for a serious approximation using the expansion of spline[x] in powers of $(x - b)$ for $b = 2$:

To the right of $x = 1$, spline[x] = $g[x]$. Taylor's formula tells you that expanding spline[x] in powers of $(x - 2)$ is the same as expanding $g[x]$ in powers of $(x - 2)$.

In[9]:=
```
b = 2; Clear[approx,x]
approx[x_] = Normal[Series[g[x],{x,b,12}]];
Plot[{spline[x],approx[x]},{x,b - 2.5,b + 2.5},
PlotStyle->{{Thickness[0.02],Blue},
{Thickness[0.01],Red}},AxesLabel->{"x",""},
Epilog->{{Red,PointSize[0.04],Point[{1,E}]},
Text["knot",{1,E},{1,-2}],
{Blue,PointSize[0.04],Point[{b,0}]},
Text["expansion point",{b,0},{0,-2}]}];
```

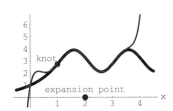

> Experiment by using more of the expansion in powers of $(x - 2)$. Do you think there's a barrier anywhere? If so, where? What do you think is the interval of convergence of the expansion of spline[x] in powers of $(x - 2)$?

G.5.b)
> Do you think that knots of splines are always barriers? Why or why not?

■ G.6) Dosing

G.6.a.i) A drug in doses of 1.5 grams is to be prescribed to a college-age woman weighing 120 pounds. She is in good physical condition. One dose of this drug results in a body fluid concentration of $(1.5/B)\,e^{-0.07t}$ grams per liter t hours after the dose. (Here B measures her body fluids in liters.) She is to take this drug every p hours, where p is set so that the concentration of this drug before each new dose never exceeds 0.032 grams per liter of her body fluids.

> Your job is to set the smallest safe value of p.

G.6.a.ii) A drug in doses of G grams is to be prescribed to a college-age woman weighing 120 pounds. She is in good physical condition. One dose of this drug results in a body fluid concentration of $(G/B)\,e^{-0.07t}$ grams per liter t hours after the dose. (Here B measures her body fluids in liters.) She is to take this drug every p hours, where

p is set so that the concentration of this drug before each new dose never exceeds 0.032 grams per liter of her body fluids.

> Your job is to give a plot of the minimum safe dose spacing as a function of the dose size G.

G.6.b) A drug in doses of G grams per dose is to be prescribed to a college-age man weighing 163 pounds. He is in good physical condition. One dose of this drug results in a body fluid concentration of $(G/B)\,e^{-0.05t}$ grams per liter t hours after the dose. (Here B measures his body fluids in liters.) The maximum safe concentration of this drug in his body fluids is 0.12 grams per liter. It has been decided to administer the drug every eight hours.

> You job is to determine the maximum safe value of the dose size G.

■ G.7) Some more uses of the expansion

$$\frac{1}{1-x} = 1 + x + x^2 + x^3 + x^4 + \cdots$$

G.7.a) Associated with each rubber ball is a bounce coefficient b. When the ball is dropped from a height h, it bounces back to a height of $b\,h$. Suppose that the ball is dropped from an initial height h, and then is allowed to bounce forever.

> Use the expansion of $1/(1-x)$ in powers of x to help come up with a clean formula that measures, in terms of b and h, the total up-down distance the ball travels in all of its bouncing.

G.7.b) The government injects \$1 billion into the economy. Each recipient spends 90% of the dollars received. In turn, the secondary recipients spend 90% of the dollars they receive, etc.

> Use the expansion of $1/(1-x)$ in powers of x to help come up with a clean measurement of the total spending that results from the \$1 billion injection.

■ G.8) More adventures of the lab pest Calculus Cal

This problem appears only in the electronic version.

■ G.9) $\sqrt{x}$ and $\log[x]$

G.9.a) Neither $\sqrt{x}$ nor $\log[x]$ has an expansion in powers of x. But if $b > 0$, then both functions have expansions in powers of $(x - b)$.

> How do you account for this?

G.9.b)
> On what interval are you guaranteed that the expansion of $\log[x]$ in powers of $(x - 2)$ converges to $\log[x]$? Illustrate with plots.

G.9.c.i)
> Go with $b > 0$. What is the expansion of $\log[x]$ in powers of $(x - b)$?
>
> On what interval are you guaranteed that the expansion of $\log[x]$ in powers of $(x - b)$ converges to $\log[x]$?

G.9.c.ii)
> Again go with $b > 0$. What is the only difference between the expansion of $\log[x]$ in powers of $(x - b)$ and the expansion of $\log[x/b]$ in powers of $(x - b)$? What property of the logarithm explains the difference?

■ G.10) Impossibilities and outrages★

For one reason or another, each of the following problems is impossible to do, or otherwise outrageous. You give the objection.

G.10.a)
> Plot $\tan[x] = \sin[x]/\cos[x]$ and enough of its expansion in powers of x on the same axes so that the two plots share the same ink on $[-\pi, \pi]$. You give the objection.

G.10.b)
> Come up with positive integer k such that
> $$1 + x + \frac{x^2}{2!} + \frac{x^3}{3!} + \frac{x^4}{4!} + \cdots + \frac{x^k}{k!}$$
> is a reasonably good approximation to e^x for all values of x. You give the objection.

> G.10.c) Plot $\sqrt{1+x}$ and enough of its expansion in powers of x on the same axes, so that the two plots share the same ink on $[-2, 2]$. You give the objection.

■ G.11) Infinite sums, decimals, and expansions

G.11.a.i) **The controversy about convergence and infinite sums**

When some one says
$$\frac{1}{1-x} = 1 + x + x^2 + x^3 + x^4 + \cdots + x^k + \cdots \qquad \text{for } -1 < x < 1,$$
the meaning of this statement is subject to a controversy that raged in mathematical circles for years and still raises its head every so often even now.

One group, including Euler, said: "This is an actual infinite sum."

The second group, including Zeno, said: "Infinite sums exist only in theory. In practice, there is no such thing as an infinite sum. Saying that
$$\frac{1}{1-x} = 1 + x + x^2 + x^3 + x^4 + \cdots + x^k + \cdots \qquad \text{for } -1 < x < 1$$
really means that, by using enough of the expansion of $1/(1-x)$ in powers of x, you can get as close to $1/(1-x)$ as you like."

The third group, including Bolzano and Weierstrass, said: "The first two groups really have nothing to fight about. You just define the infinite sum
$$\frac{1}{1-x} = 1 + x + x^2 + x^3 + x^4 + \cdots + x^k + \cdots$$
to mean that given an x with $-1 < x < 1$ and a tolerance t, then you can find a power m so that
$$\left| \frac{1}{1-x} - \left(1 + x + x^2 + x^3 + \cdots + x^m\right) \right| < t."$$

> Which group are you in, and why?

G.11.a.ii) By and large, twentieth century mathematics has adopted the position of the third group. Some mathematicians who feel the very strong influence of computers on mathematics today are converting to the second group. When you go on into advanced mathematics, you will probably be taught that the position of the third group is the position you must adopt.

> Are you happy with this, or do you regard the third group's position as nothing more than a play on words?

As always in Calculus&*Mathematica*, assign yourself the freedom of expressing your own feelings.

G.11.b) This problem appears only in the electronic version.

■ G.12) Point fit with polynomials, expansions, and convergence intervals

This problem appears only in the electronic version.

LESSON 4.06

Power Series

Basics

■ B.1) Functions defined by power series

B.1.a) What is a power series? Why are power series big deals?

Answer: Any expansion of a function in powers of $(x - b)$ is a power series. Sometimes you use power series to come up with the expansion of a function without having your hands on a formula for the function. In these cases, the function is defined by its expansion. For instance, if the power series you see is
$$1 + x + x^2 + x^3 + x^4 + \cdots + x^k + \cdots,$$
then you recognize this power series as the expansion of
$$f[x] = \frac{1}{1-x}$$
in powers of x. On the other hand, if the power series you see is
$$1 + x + \frac{1}{2^2}x^2 + \frac{1}{3^2}x^3 + \frac{1}{4^2}x^4 + \cdots + \frac{1}{k^2}x^k + \cdots,$$
then you recognize this power series as the expansion of a function $f[x]$, but you probably don't know a clean formula for $f[x]$. In this case, the best you can say is that $f[x]$ is defined by this power series.

B.1.b) One function that is defined by a power series is the function $f[x]$, whose expansion in powers of x is
$$1 - x + \frac{1}{2^2}x^2 - \frac{1}{3^2}x^3 + \frac{1}{4^2}x^4 + \cdots \frac{(-1)^k}{k^2}x^k + \cdots$$

Lesson 4.06 Power Series

> What information about $f[x]$ can you glean from this power series?

Answer: You can get an idea of how $f[x]$ plots out on short intervals centered at 0. Look at this:

In[1]:=
```
Clear[expan,x,m,k]
expan[x_,m_] = 1 + Sum[((-1)^k x^k)/k^2,{k,1,m}];
```

The expansion of $f[x]$ in powers of x through the x^8 term is:

In[2]:=
```
expan[x,8]
```
Out[2]=

$$1 - x + \frac{x^2}{4} - \frac{x^3}{9} + \frac{x^4}{16} - \frac{x^5}{25} + \frac{x^6}{36} - \frac{x^7}{49} + \frac{x^8}{64}$$

The expansion of $f[x]$ in powers of x through the x^9 term is:

In[3]:=
```
expan[x,9]
```
Out[3]=

$$1 - x + \frac{x^2}{4} - \frac{x^3}{9} + \frac{x^4}{16} - \frac{x^5}{25} + \frac{x^6}{36} - \frac{x^7}{49} + \frac{x^8}{64} - \frac{x^9}{81}$$

Now look at these plots of the expansions of $f[x]$ through the x^8 and x^9 terms:

In[4]:=
```
Plot[{expan[x,8],expan[x,9]},
 {x,-1.5,1.5},PlotStyle->
 {{Thickness[0.02],Blue},
 {Thickness[0.01],Red}},
 AxesLabel->{"x",""}];
```

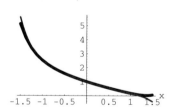

And look at the plots of the expansions of $f[x]$ through the x^{12} and x^{13} terms:

In[5]:=
```
Plot[{expan[x,12],expan[x,13]},
 {x,-1.5,1.5},PlotStyle->
 {{Thickness[0.02],Blue},
 {Thickness[0.01],Red}},
 AxesLabel->{"x",""}];
```

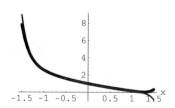

Fairly strong evidence of barriers near $x = -1$ and $x = 1$. It seems fairly safe to say that a reasonably trustworthy plot of $f[x]$ is:

```
In[6]:=
    Plot[expan[x,13],{x,-0.8,0.8},
    PlotStyle->{{Thickness[0.01],Blue}},
    AxesLabel->{"x","f[x]"}];
```

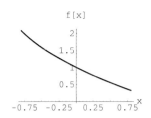

Think of it: All this information with no formula for $f[x]$.

B.1.c) Try to get a reasonably trustworthy plot of the function $f[x]$ defined by the power series

$$1 - \frac{x^2}{\sqrt{2!}} + \frac{x^4}{\sqrt{4!}} - \frac{x^6}{\sqrt{6!}} + \frac{x^8}{\sqrt{8!}} + \cdots + \frac{(-1)^k x^{2k}}{\sqrt{(2k)!}} + \cdots$$

Answer: Enter the partial expansion:

```
In[7]:=
    Clear[expan,x,m,k]
    expan[x_,m_] = 1 + Sum[(-1)^k x^(2 k)/Sqrt[(2 k)!],{k,1,m}];
```

Here is the expansion through the x^{14} term:

```
In[8]:=
    expan[x,7]
```

Out[8]=

$$1 - \frac{x^2}{\text{Sqrt}[2]} + \frac{x^4}{2\,\text{Sqrt}[6]} - \frac{x^6}{12\,\text{Sqrt}[5]} + \frac{x^8}{24\,\text{Sqrt}[70]} - \frac{x^{10}}{720\,\text{Sqrt}[7]} + \frac{x^{12}}{1440\,\text{Sqrt}[231]} - \frac{x^{14}}{10080\,\text{Sqrt}[858]}$$

And the x^{16} term:

```
In[9]:=
    expan[x,8]
```

Out[9]=

$$1 - \frac{x^2}{\text{Sqrt}[2]} + \frac{x^4}{2\,\text{Sqrt}[6]} - \frac{x^6}{12\,\text{Sqrt}[5]} + \frac{x^8}{24\,\text{Sqrt}[70]} - \frac{x^{10}}{720\,\text{Sqrt}[7]} + \frac{x^{12}}{1440\,\text{Sqrt}[231]} - \frac{x^{14}}{10080\,\text{Sqrt}[858]} + \frac{x^{16}}{120960\,\text{Sqrt}[1430]}$$

The important thing to note here is that $\text{expan}[x,7]$ and $\text{expan}[x,8]$ are not the same. Here are their plots on $[-4,4]$:

Lesson 4.06 Power Series

```
In[10]:=
    r = 4;
    Plot[{expan[x,7],expan[x,8]},{x,-r,r},
    PlotStyle->{{Thickness[0.02],Blue},
    {Thickness[0.01],Red}},
    AxesLabel->{"x",""}];
```

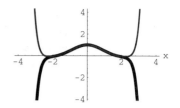

The plots share a lot of ink on the interval $[-2, 2]$. Take a closer look:

```
In[11]:=
    r = 2;
    Plot[{expan[x,7],expan[x,8]},{x,-r,r},
    PlotStyle->{{Thickness[0.02],Blue},
    {Thickness[0.01],Red}},
    PlotRange->All,
    AxesLabel->{"x",""}];
```

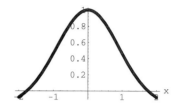

The plot tells everything. This is very strong evidence that a trustworthy plot of $f[x]$ is:

```
In[12]:=
    r = 2;
    Plot[expan[x,8],{x,-r,r},
    PlotStyle->{{Thickness[0.02],Blue}},
    PlotRange->All,
    AxesLabel->{"x","f[x]"}];
```

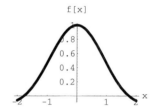

You can go after an accurate plot of $f[x]$ over longer intervals than this by using more of the expansion, but unless you've got a real hot rod machine, you'll wait a long time for it. This is the trouble with power series; they often converge too slowly to be of any use for a plot on a long interval.

B.1.d) What do professional software writers do with functions defined by power series?

Answer: This is a big league question. Professional software writers use power series the way biologists use DNA. Professional software writers take the data contained in the power series and feed them into advanced approximation schemes to come up with incredibly accurate approximations of a function defined by a given power series. This is true for everyday functions like $f[x] = \sin[x]$. For instance, when *Mathematica* evaluates $\sin[x]$, it does not make measurements on triangles. *Mathematica* takes the expansion of $\sin[x]$ in powers of x,

$$x - \frac{x^3}{3!} + \frac{x^5}{5!} - \frac{x^7}{7!} + \frac{x^9}{9!} + \cdots \frac{(-1)^k \, x^{(2k+1)}}{(2\,k+1)!} + \cdots \, ;$$

then it extracts the data it needs from the expansion and creates its own efficient, accurate approximation of sin[x]. The actual process is both fascinating and technical. If this catches your interest, you're a good candidate for going on to advanced work in mathematics.

■ B.2) Functions defined by power series via differential equations

B.2.a) You can say that e^x is defined by the differential equation

$$y'[x] - y[x] = 0 \quad \text{with } y[0] = 1:$$

In[13]:=
```
Clear[y,x,Derivative]
DSolve[{y'[x] - y[x] == 0,y[0] == 1},y[x],x]
```

Out[13]=
```
{{y[x] -> E^x}}
```

You can come up with the power series

$$1 + x + \frac{x^2}{2!} + \frac{x^3}{3!} + \cdots + \frac{x^k}{k!} + \cdots,$$

which is the expansion of e^x in powers of x, directly from the idea of order of contact, or from Taylor's formula. It is also possible to come up with this expansion directly from the differential equation $y'[x] - y[x] = 0$ with $y[0] = 1$.

> **How?**

Answer: You use the method of undetermined coefficients.

Here's how it goes: The differential equation is

$$y'[x] - y[x] = 0 \quad \text{with } y[0] = 1.$$

Here is the expansion of $y[x]$ through the x^8 term with most of the coefficients undetermined:

In[14]:=
```
Clear[a,x,k,prelimexpan]; m = 8;
prelimexpan[x_] = (Sum[a[k] x^k,{k,0,m}] + O[x]^(m + 1))/.a[0]->1
```

Out[14]=
```
1 + a[1] x + a[2] x  + a[3] x  + a[4] x  + a[5] x  + a[6] x  + a[7] x  +
                    2          3          4          5          6          7

a[8] x  + O[x]
      8        9
```

The reason you take $a[0] = 1$ is that by Taylor's formula $1 = y[0] = a[0]$. The differential equation is $y'[x] - y[x] = 0$; replace $y[x]$ by the partial expansion:

Lesson 4.06 Power Series

In[15]:=
```
left = prelimexpan'[x] - prelimexpan[x]
```

Out[15]=

$(-1 + a[1]) + (-a[1] + 2\,a[2])\,x + (-a[2] + 3\,a[3])\,x^2 + (-a[3] + 4\,a[4])\,x^3 +$
$(-a[4] + 5\,a[5])\,x^4 + (-a[5] + 6\,a[6])\,x^5 + (-a[6] + 7\,a[7])\,x^6 +$
$(-a[7] + 8\,a[8])\,x^7 + O[x]^8$

In[16]:=
```
right = 0
```

Out[16]=
0

Set the left side equal to the right side and equate the coefficients of like powers of x. On the right, all the coefficients of all the powers of x must be 0. This gives:

$$-1 + a[1] = 0,$$
$$(-a[1] + 2\,a[2]) = 0,$$
$$(-a[2] + 3\,a[3]) = 0,$$

etc. *Mathematica* will set these equations up for you:

In[17]:=
```
coeffeqns = LogicalExpand[left == right]
```

Out[17]=
```
-1 + a[1]  == 0 && -a[1] + 2 a[2] == 0 &&
  -a[2] + 3 a[3] == 0 &&
  -a[3] + 4 a[4] == 0 &&
  -a[4] + 5 a[5] == 0 &&
  -a[5] + 6 a[6] == 0 &&
  -a[6] + 7 a[7] == 0 && -a[7] + 8 a[8] == 0
```

Solve these equations:

In[18]:=
```
coeffsolved = Solve[coeffeqns]
```

Out[18]=

$\{\{a[1] \to 1,\ a[2] \to \frac{1}{2},\ a[3] \to \frac{1}{6},\ a[4] \to \frac{1}{24},\ a[5] \to \frac{1}{120},\ a[6] \to \frac{1}{720},$
$a[7] \to \frac{1}{5040},\ a[8] \to \frac{1}{40320}\}\}$

Substitute these into the preliminary partial expansion of $y[x]$:

In[19]:=
```
Clear[expan]
expan[x_] = prelimexpan[x]/.coeffsolved[[1]]
```

Out[19]=
$$1 + x + \frac{x^2}{2} + \frac{x^3}{6} + \frac{x^4}{24} + \frac{x^5}{120} + \frac{x^6}{720} + \frac{x^7}{5040} + \frac{x^8}{40320} + O[x]^9$$

Chop the error term to get the expansion through the eighth degree term:

In[20]:=
```
Clear[expan8]; expan8[x_] = Normal[expan[x]]
```

Out[20]=
$$1 + x + \frac{x^2}{2} + \frac{x^3}{6} + \frac{x^4}{24} + \frac{x^5}{120} + \frac{x^6}{720} + \frac{x^7}{5040} + \frac{x^8}{40320}$$

To get more of the expansion, just copy, paste, and run with a larger m:

In[21]:=
```
Clear[x,k,prelimexpan]; m = 13;
prelimexpan[x_] = (Sum[a[k] x^k,{k,0,m}] + O[x]^(m + 1))/.a[0]->1;
left = prelimexpan'[x] - prelimexpan[x]; right = 0;
coeffeqns = LogicalExpand[left == right];
coeffsolved = Solve[coeffeqns]; Clear[expan13]
expan13[x_] = Normal[prelimexpan[x]/.coeffsolved[[1]]]
```

Out[21]=
$$1 + x + \frac{x^2}{2} + \frac{x^3}{6} + \frac{x^4}{24} + \frac{x^5}{120} + \frac{x^6}{720} + \frac{x^7}{5040} + \frac{x^8}{40320} + \frac{x^9}{362880} + \frac{x^{10}}{3628800} +$$
$$\frac{x^{11}}{39916800} + \frac{x^{12}}{479001600} + \frac{x^{13}}{6227020800}$$

This is the same as:

In[22]:=
```
Normal[Series[E^x,{x,0,13}]]
```

Out[22]=
$$1 + x + \frac{x^2}{2} + \frac{x^3}{6} + \frac{x^4}{24} + \frac{x^5}{120} + \frac{x^6}{720} + \frac{x^7}{5040} + \frac{x^8}{40320} + \frac{x^9}{362880} + \frac{x^{10}}{3628800} +$$
$$\frac{x^{11}}{39916800} + \frac{x^{12}}{479001600} + \frac{x^{13}}{6227020800}$$

The method works.

B.2.b.i) The function BesselJ$[0, x]$ is defined to be the solution of the differential equation

$$y''[x] + \frac{y'[x]}{x} + y[x] = 0 \quad \text{with } y[0] = 1 \text{ and } y'[0] = 0:$$

> Use the method of undetermined coefficients to come up with the expansion of BesselJ$[0, x]$ through the x^{16} term.

Lesson 4.06 Power Series

> Plot this part of the expansion of BesselJ$[0, x]$ and the actual function BesselJ$[0, x]$ for $-7.5 \leq x \leq 7.5$.

Answer: The differential equation is

$$y''[x] + \frac{y'[x]}{x} + y[x] = 0 \qquad \text{with } y[0] = 1 \text{ and } y'[0] = 0.$$

Here is the expansion of $y[x]$ through the x^{16} term with most of the coefficients undetermined:

In[23]:=
```
Clear[x,a,k,prelimexpan]; m = 16;
prelimexpan[x_] = (Sum[a[k] x^k,{k,0,m}] + O[x]^(m + 1))/.{a[0]->1,a[1]->0}
```

Out[23]=

$1 + a[2] x^2 + a[3] x^3 + a[4] x^4 + a[5] x^5 + a[6] x^6 + a[7] x^7 + a[8] x^8 +$
$a[9] x^9 + a[10] x^{10} + a[11] x^{11} + a[12] x^{12} + a[13] x^{13} + a[14] x^{14} +$
$a[15] x^{15} + a[16] x^{16} + O[x]^{17}$

The reason you take $a[0] = 1$ and $a[1] = 0$ is that by Taylor's formula $1 = y[0] = a[0]$ and $0 = y'[0] = a[1]$. The differential equation is

$$y''[x] + \frac{y[x]}{x} + y[x] = 0;$$

replace $y[x]$ by the partial expansion:

In[24]:=
```
left = prelimexpan''[x] + prelimexpan'[x]/x + prelimexpan[x]
```

Out[24]=

$(1 + 4\, a[2]) + 9\, a[3]\, x + (a[2] + 16\, a[4])\, x^2 + (a[3] + 25\, a[5])\, x^3 +$
$(a[4] + 36\, a[6])\, x^4 + (a[5] + 49\, a[7])\, x^5 + (a[6] + 64\, a[8])\, x^6 +$
$(a[7] + 81\, a[9])\, x^7 + (a[8] + 100\, a[10])\, x^8 + (a[9] + 121\, a[11])\, x^9 +$
$(a[10] + 144\, a[12])\, x^{10} + (a[11] + 169\, a[13])\, x^{11} +$
$(a[12] + 196\, a[14])\, x^{12} + (a[13] + 225\, a[15])\, x^{13} +$
$(a[14] + 256\, a[16])\, x^{14} + O[x]^{15}$

In[25]:=
```
right = 0
```

Out[25]=
0

Set the left side equal to the right side and equate the coefficients of like powers of x. On the right, all the coefficients of the powers of x must be 0. This gives:

$$1 + 4\,a[2] = 0,$$
$$9\,a[3] = 0,$$
$$a[2] + 16\,a[4] = 0, \text{ etc.}$$

Mathematica will set these equations up for you:

In[26]:=
  `coeffeqns = LogicalExpand[left == right]`

Out[26]=
  1 + 4 a[2] == 0 && 9 a[3] == 0 &&
    a[2] + 16 a[4] == 0 &&
    a[3] + 25 a[5] == 0 &&
    a[4] + 36 a[6] == 0 &&
    a[5] + 49 a[7] == 0 &&
    a[6] + 64 a[8] == 0 &&
    a[7] + 81 a[9] == 0 &&
    a[8] + 100 a[10] == 0 &&
    a[9] + 121 a[11] == 0 &&
    a[10] + 144 a[12] == 0 &&
    a[11] + 169 a[13] == 0 &&
    a[12] + 196 a[14] == 0 &&
    a[13] + 225 a[15] == 0 &&
    a[14] + 256 a[16] == 0

Solve these equations:

In[27]:=
  `coeffsolved = Solve[coeffeqns]`

Out[27]=
  {{a[2] -> -($\frac{1}{4}$), a[3] -> 0, a[4] -> $\frac{1}{64}$, a[5] -> 0, a[6] -> -($\frac{1}{2304}$), a[7] -> 0,
    a[8] -> $\frac{1}{147456}$, a[9] -> 0, a[10] -> -($\frac{1}{14745600}$), a[11] -> 0,
    a[12] -> $\frac{1}{2123366400}$, a[13] -> 0, a[14] -> -($\frac{1}{416179814400}$), a[15] -> 0,
    a[16] -> $\frac{1}{106542032486400}$}}

Substitute these into the preliminary partial expansion of y:

In[28]:=
  `Clear[expan]`
  `expan[x_] = prelimexpan[x]/.coeffsolved[[1]]`

Out[28]=
$$1 - \frac{x^2}{4} + \frac{x^4}{64} - \frac{x^6}{2304} + \frac{x^8}{147456} - \frac{x^{10}}{14745600} + \frac{x^{12}}{2123366400} - \frac{x^{14}}{416179814400} + \frac{x^{16}}{106542032486400} + O[x]^{17}$$

Chop the error term to get the expansion through the sixteenth degree term:

In[29]:=
```
Clear[expan16]
expan16[x_] = Normal[expan[x]]
```

Out[29]=
$$1 - \frac{x^2}{4} + \frac{x^4}{64} - \frac{x^6}{2304} + \frac{x^8}{147456} - \frac{x^{10}}{14745600} + \frac{x^{12}}{2123366400} - \frac{x^{14}}{416179814400} + \frac{x^{16}}{106542032486400}$$

Compare:

In[30]:=
```
Normal[Series[BesselJ[0,x],{x,0,16}]]
```

Out[30]=
$$1 - \frac{x^2}{4} + \frac{x^4}{64} - \frac{x^6}{2304} + \frac{x^8}{147456} - \frac{x^{10}}{14745600} + \frac{x^{12}}{2123366400} - \frac{x^{14}}{416179814400} + \frac{x^{16}}{106542032486400}$$

Perfecto. Here comes the plot:

In[31]:=
```
Plot[{BesselJ[0,x],expan16[x]},
 {x,-7.5,7.5},PlotStyle->
 {{Thickness[0.02],Blue},
 {Thickness[0.01],Red}},
 AxesLabel->{"x",""}];
```

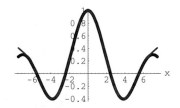

It's everything you could have hoped for.

B.2.b.ii) What's the main difference between this method of coming up with the expansion of BesselJ$[0, x]$ in powers of x and the way you got expansions in previous lessons?

Answer: The difference is huge. In previous lessons, you used clean formulas and derivatives and order of contact. Here you have no clean formula for BesselJ[0, x]. Despite this, you were able to come up with as much as you want of the expansion of BesselJ[0, x]. Just do it:

In[32]:=
```
Clear[x,k,prelimexpan]; m = 24;
prelimexpan[x_] = (Sum[a[k] x^k,{k,0,m}] + O[x]^(m + 1))/.{a[0]->1,a[1]->0};
left = prelimexpan''[x] + prelimexpan'[x]/x + prelimexpan[x];
right = 0; coeffeqns = LogicalExpand[left == right];
coeffsolved = Solve[coeffeqns]; Clear[expan]
expan[x_] = prelimexpan[x]/.coeffsolved[[1]];
expan16[x_] = Normal[expan[x]]
```

Out[32]=
$$1 - \frac{x^2}{4} + \frac{x^4}{64} - \frac{x^6}{2304} + \frac{x^8}{147456} - \frac{x^{10}}{14745600} + \frac{x^{12}}{2123366400} - \frac{x^{14}}{416179814400} +$$
$$\frac{x^{16}}{106542032486400} - \frac{x^{18}}{34519618525593600} + \frac{x^{20}}{13807847410237440000} -$$
$$\frac{x^{22}}{6682998146554920960000} + \frac{x^{24}}{384940693241563447296000}$$

Compare:

In[33]:=
```
Normal[Series[BesselJ[0,x],{x,0,m}]]
```

Out[33]=
$$1 - \frac{x^2}{4} + \frac{x^4}{64} - \frac{x^6}{2304} + \frac{x^8}{147456} - \frac{x^{10}}{14745600} + \frac{x^{12}}{2123366400} - \frac{x^{14}}{416179814400} +$$
$$\frac{x^{16}}{106542032486400} - \frac{x^{18}}{34519618525593600} + \frac{x^{20}}{13807847410237440000} -$$
$$\frac{x^{22}}{6682998146554920960000} + \frac{x^{24}}{384940693241563447296000}$$

On the money.

B.2.c.i) Come up with the expansion in powers of x through the x^{10} term of the solution to the differential equation

$$y''[x] + \sin[x]\, y'[x] + (1 + x^2)\, y[x] = e^{-x} \qquad \text{with } y[0] = 1 \text{ and } y'[0] = -0.7.$$

Answer: The differential equation is $y''[x] + \sin[x]\, y'[x] + (1 + x^2)\, y[x] = e^{-x}$, with $y[0] = 1$ and $y'[0] = -0.7$. Here is the expansion of $y[x]$ through the x^{10} term with most of the coefficients undetermined:

Lesson 4.06 Power Series

$In[34]:=$
```
Clear[x,k,a,prelimexpan]; m = 10;
prelimexpan[x_] = (Sum[a[k] x^k,{k,0,m}] + O[x]^(m + 1))/.{a[0]->1,a[1]->-0.7}
```

$Out[34]=$

$1 - 0.7 x + a[2] x^2 + a[3] x^3 + a[4] x^4 + a[5] x^5 + a[6] x^6 + a[7] x^7 + a[8] x^8 + a[9] x^9 + a[10] x^{10} + O[x]^{11}$

The reason you take $a[0] = 1$ and $a[1] = -0.7$ is that Taylor's formula says

$$1 = y[0] = a[0]$$

and

$$-0.7 = y'[0] = a[1].$$

The differential equation is

$$y''[x] + \sin[x]\, y'[x] + (1 + x^2)\, y[x] = e^{-x} \qquad \text{with } y[0] = 1 \text{ and } y'[0] = -0.7.$$

Replace $y[x]$, $\sin[x]$, and e^{-x} by their partial expansions:

$In[35]:=$
```
sinexpan = Series[Sin[x],{x,0,m}];
eexpan = Series[E^(-x),{x,0,m}];
left = prelimexpan''[x] + sinexpan prelimexpan'[x] + (1 + x^2) prelimexpan[x]
```

$Out[35]=$

$(1 + 2\, a[2]) + (-1.4 + 6\, a[3])\, x + (1 + 3\, a[2] + 12\, a[4])\, x^2 +$

$(-0.583333 + 4\, a[3] + 20\, a[5])\, x^3 + (\dfrac{2\, a[2]}{3} + 5\, a[4] + 30\, a[6])\, x^4 +$

$(-0.00583333 + \dfrac{a[3]}{2} + 6\, a[5] + 42\, a[7])\, x^5 +$

$(\dfrac{a[2]}{60} + \dfrac{a[4]}{3} + 7\, a[6] + 56\, a[8])\, x^6 +$

$(0.000138889 + \dfrac{a[3]}{40} + \dfrac{a[5]}{6} + 8\, a[7] + 72\, a[9])\, x^7 +$

$(\dfrac{-a[2]}{2520} + \dfrac{a[4]}{30} + 9\, a[8] + 90\, a[10])\, x^8 + O[x]^9$

$In[36]:=$
```
right = eexpan
```

$Out[36]=$

$1 - x + \dfrac{x^2}{2} - \dfrac{x^3}{6} + \dfrac{x^4}{24} - \dfrac{x^5}{120} + \dfrac{x^6}{720} - \dfrac{x^7}{5040} + \dfrac{x^8}{40320} - \dfrac{x^9}{362880} + \dfrac{x^{10}}{3628800} + O[x]^{11}$

Set the left side equal to the right side and equate the coefficients of like powers of x. On the right, all the coefficients of the powers of x must be 0. This gives:

$$1 + 2\,a[2] = 1,$$
$$-1.4 + 6\,a[3] = -1,$$
$$1 - a[2] + 12\,a[4] = \frac{1}{2},$$

etc. *Mathematica* will set these equations up for you:

In[37]:=
   coeffeqns = LogicalExpand[left == right]

Out[37]=

2 a[2] == 0 && -0.4 + 6 a[3] == 0 && $\frac{1}{2}$ + 3 a[2] + 12 a[4] == 0 &&

-0.416667 + 4 a[3] + 20 a[5] == 0 &&

-($\frac{1}{24}$) + $\frac{2\,a[2]}{3}$ + 5 a[4] + 30 a[6] == 0 &&

0.0025 + $\frac{a[3]}{2}$ + 6 a[5] + 42 a[7] == 0 &&

-($\frac{1}{720}$) + $\frac{a[2]}{60}$ + $\frac{a[4]}{3}$ + 7 a[6] + 56 a[8] == 0 &&

0.000337302 + $\frac{a[3]}{40}$ + $\frac{a[5]}{6}$ + 8 a[7] + 72 a[9] == 0 &&

-($\frac{1}{40320}$) - $\frac{a[2]}{2520}$ + $\frac{a[4]}{30}$ + 9 a[8] + 90 a[10] == 0

Solve these equations, chopping out the negligible terms:

In[38]:=
   coeffsolved = Chop[Solve[coeffeqns]]

Out[38]=
   {{a[2] -> 0, a[3] -> 0.0666667, a[4] -> -0.0416667, a[5] -> 0.0075,
     a[6] -> 0.00833333, a[7] -> -0.0019246, a[8] -> -0.000768849,
     a[9] -> 0.000168651, a[10] -> 0.0000925926}}

Substitute these into the preliminary partial expansion of y and chop the error term:

In[39]:=
   Clear[expan]
   expan[x_] = prelimexpan[x]/.coeffsolved[[1]]

Out[39]=
   $1 - 0.7\,x + 0.0666667\,x^3 - 0.0416667\,x^4 + 0.0075\,x^5 + 0.00833333\,x^6 - 0.0019246\,x^7 - 0.000768849\,x^8 + 0.000168651\,x^9 + 0.0000925926\,x^{10} + O[x]^{11}$

Get rid of the error term to get the expansion through the tenth degree term:

Lesson 4.06 Power Series

In[40]:=
```
Clear[expan10]
expan10[x_] = Normal[expan[x]]
```

Out[40]=

$1 - 0.7 x + 0.0666667 x^3 - 0.0416667 x^4 + 0.0075 x^5 + 0.00833333 x^6 - 0.0019246 x^7 - 0.000768849 x^8 + 0.000168651 x^9 + 0.0000925926 x^{10}$

Not too bad.

B.2.c.ii) Use power series to give what you feel to be a trustworthy plot of the solution of

$$y''[x] + \sin[x]\, y'[x] + (1 + x^2)\, y[x] = e^{-x} \quad \text{with } y[0] = 1 \text{ and } y'[0] = -0.7.$$

Check your plot with NDSolve.

Answer: Copy, paste, and edit the answer from part B.2.c.i) above.

In[41]:=
```
Clear[x,k,a,prelimexpan]; m = 16;
prelimexpan[x_] = (Sum[a[k] x^k,{k,0,m}] + O[x]^(m + 1))/.{a[0]->1,a[1]->-0.7};
sinexpan = Series[Sin[x],{x,0,m}]; eexpan = Series[E^(-x),{x,0,m}];
left = prelimexpan''[x] + sinexpan prelimexpan'[x] + (1 + x^2) prelimexpan[x];
right = eexpan; coeffeqns = LogicalExpand[left == right];
coeffsolved = Chop[Solve[coeffeqns]]; Clear[expan16]
expan16[x_] = Normal[prelimexpan[x]/.coeffsolved[[1]]]
```

Out[41]=

$1 - 0.7 x + 0.0666667 x^3 - 0.0416667 x^4 + 0.0075 x^5 + 0.00833333 x^6 - 0.0019246 x^7 - 0.000768849 x^8 + 0.000168651 x^9 + 0.0000925926 x^{10} + 0.0000207356 x^{11} - 0.0000130626 x^{12} + 2.89916\, 10^{-6} x^{13} + 1.61088\, 10^{-6} x^{14} - 3.48991\, 10^{-7} x^{15} - 1.92934\, 10^{-7} x^{16}$

Note the plus sign in the x^{14} term.

In[42]:=
```
Clear[expan14]
expan14[x_] = Normal[Series[expan16[x],{x,0,14}]]
```

Out[42]=

$1 - 0.7 x + 0.0666667 x^3 - 0.0416667 x^4 + 0.0075 x^5 + 0.00833333 x^6 - 0.0019246 x^7 - 0.000768849 x^8 + 0.000168651 x^9 + 0.0000925926 x^{10} + 0.0000207356 x^{11} - 0.0000130626 x^{12} + 2.89916\, 10^{-6} x^{13} + 1.61088\, 10^{-6} x^{14}$

Now plot:

In[43]:=
```
r = 4;
Plot[{expan14[x],expan16[x]},{x,-r,r},
PlotStyle->{{Thickness[0.02],Blue},
{Thickness[0.01],Red}},
AxesLabel->{"x",""}];
```

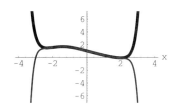

Cut back the interval:

In[44]:=
```
r = 2;
Plot[{expan14[x],expan16[x]},{x,-r,r},
PlotStyle->{{Thickness[0.02],Blue},
{Thickness[0.01],Red}},
AxesLabel->{"x",""}];
```

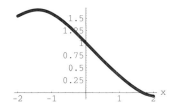

A trustworthy plot of the solution of this differential equation is:

In[45]:=
```
solutionplot = Plot[expan16[x],{x,-2,2},
PlotStyle->{{Thickness[0.02],Blue}},
AxesLabel->{"x","y[x]"}];
```

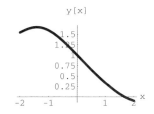

Compare to what you get from NDSolve:

In[46]:=
```
a = -2; c = 2; Clear[solution,x,y,fakey]
solution = NDSolve[{y''[x] + Sin[x] y'[x] + (1 + x^2) y[x] == E^(-x),
y[0] == 1, y'[0] == - 0.7},y[x],{x,a,c}];
fakey[x_] = y[x]/.solution[[1]];
Mathematicaplot = Plot[fakey[x],{x,a,c},
PlotStyle->{{Red,Thickness[0.01]}},PlotRange->All,DisplayFunction->Identity];
```

In[47]:=
```
Show[solutionplot,Mathematicaplot,
DisplayFunction->$DisplayFunction];
```

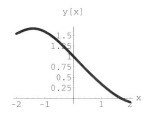

Just fine; thank you.

B.2.c.iii) Which should you use, power series or NDSolve to come up with a trustworthy plot of
$$y''[x] + \sin[x]\,y'[x] + (1+x^2)\,y[x] = e^{-x}$$
with $y[0] = 1$ and $y'[0] = -0.7$ for $-5 \le x \le 5$?

Answer: You don't have a heck of a lot of choice. Even when you ran out the power series to the x^{16} term, you got trustworthy plots for $-2 \le x \le 2$ only. To use power series to go after a trustworthy plot on the big interval $-5 \le x \le 5$, you'd have to run the power series way out to very high-degree terms. The resulting calculations would be monumental. For the big interval $-5 \le x \le 5$, your best bet is to go with NDSolve:

In[48]:=
```
a = -5; c = 5; Clear[solution,x,y,fakey]
solution = NDSolve[{y''[x] + Sin[x] y'[x] + (1 + x^2) y[x] == E^(-x),
  y[0] == 1,y'[0] == - 0.7},y[x],{x,a,c}];
fakey[x_] = y[x]/.solution[[1]];
```

In[49]:=
```
Plot[fakey[x],{x,a,c},
  PlotStyle->{{Blue,Thickness[0.02]}},
  PlotRange->All,AxesLabel->{"x",""}];
```

B.2.c.iv) Why is the power series method of coming up with trustworthy plots of solutions to differential equations on long intervals likely to be impractical?

Answer: On long intervals the power series are likely to converge so slowly that in order to get any accuracy, you have to use a humongous number of terms. When you feed a humongous number of terms into the machine, the machine is likely to bog down. Usually, no such problem comes up when you use NDSolve.

■ B.3) Convergence intervals for functions defined by power series via differential equations

If you have a power series coming from the solution of a differential equation, you can often inspect the differential equation to determine a convergence interval for the expansion.

B.3.a) If you are dealing with
$$P[x]\,y'[x] + Q[x]\,y[x] = f[x],$$

then you are guaranteed that the convergence interval of the expansion of solution in powers of x is at least as big as the smallest of the convergence intervals of the expansions of $1/P[x], Q[x]$ and $f[x]$.

> Apply this information to the power series you get for the solution of the differential equation
> $$\left(1+x^2\right) y'[x] + \frac{1}{(1-2x)} y[x] = \cos[x] \quad \text{with } y[0]=1.$$

Answer: The convergence interval of the expansion of $1/\left(1+x^2\right)$ in powers of x is $-1 < x < 1$.

The convergence interval of the expansion of $1/(1-2x)$ in powers of x is $-1/2 < x < 1/2$.

The convergence interval of the expansion of $\cos[x]$ in powers of x is $-\infty < x < \infty$.

The smallest of these intervals is $-1/2 < x < 1/2$. This tells you that the convergence interval of the expansion of the solution of the differential equation $\left(1+x^2\right) y'[x] + (1/(1-2x)) y[x] = \cos[x]$ with $y[0]=1$ in powers of x is at least as big as $-1/2 < x < 1/2$.

B.3.b) If you are dealing with $P[x]y''[x] + Q[x]y'[x] + R[x]y[x] = f[x]$, then you are guaranteed that the convergence interval of the expansion of solution in powers of x is at least as big as the smallest of the convergence intervals of the expansions of $1/P[x], Q[x], R[x]$, and $f[x]$ in powers of x.

> Apply this information to the power series you get for the solution of the differential equation
> $$\left(1+x^2\right) y''[x] + x\, y'[x] - e^{-x} y[x] = \sin[x] \quad \text{with } y[0]=1 \text{ and } y'[0]=-\frac{1}{2}.$$

Answer: The convergence interval of the expansion of $1/\left(1+x^2\right)$ in powers of x is $-1 < x < 1$.

The convergence interval of the expansion of x in powers of x is $-\infty < x < \infty$.

The convergence interval of the expansion of e^{-x} in powers of x is $-\infty < x < \infty$.

The convergence interval of the expansion of $\sin[x]$ in powers of x is $-\infty < x < \infty$.

The smallest of these intervals is $-1 < x < 1$. This tells you that the convergence interval of the expansion of the solution of the differential equation
$$\left(1+x^2\right) y''[x] + x\, y'[x] - e^{-x} y[x] = \sin[x] \quad \text{with } y[0]=1 \text{ and } y'[0]=-\frac{1}{2}$$
in powers of x is at least as big as $-1 < x < 1$.

B.3.c) Do these rules produce the biggest possible convergence intervals?

Answer: Not always. Sometimes the actual convergence interval is much longer than the interval that these rules guarantee. For instance, in B.2) above, the differential equation was

$$y''[x] + \frac{y'[x]}{x} + y[x] = 0 \quad \text{with } y[0] = 1 \text{ and } y'[0] = 0.$$

Since $1/x$ has no expansion in powers of x (it has a gigantic singularity at $x = 0$), the rules above predict no convergence interval for the expansion of $y[x]$ in powers of x. Yet, in this case, the actual convergence interval for the expansion of $y[x]$ in powers of x is $-\infty < x < \infty$. This fact can be seen from some advanced techniques that you do not need to concern yourself with.

■ B.4) Convergence intervals for general power series

B.4.a.i) Given a specific power series, how do you try to estimate where the barriers are?

Answer: You can try eyeballing some plots.

B.4.a.ii) How do you try to come up with more definitive information than you get from eyeball estimates?

Answer: You can try to use the Power Series Convergence Principle. This principle says that if $f[x]$ is defined by a power series

$$a[0] + a[1]\, x + a[2]\, x^2 + a[3]\, x^3 + \cdots + a[k]\, x^k + \cdots,$$

where the $a[k]$'s are given constants, then you are guaranteed that the power series converges to $f[x]$ for $-R < x < R$, provided that the infinite list of individual terms

$$\{a[0], a[1]\, R, a[2]\, R^2, a[3]\, R^3, \ldots, a[k]\, R^k, \ldots\}$$

stays bounded (i.e., does not blow up to ∞ or down to $-\infty$).

B.4.b.i) Go back to the function $f[x]$ whose expansion in powers of x is

$$1 - x + \frac{x^2}{2^2} - \frac{x^3}{3^2} + \frac{x^4}{4^2} + \cdots \frac{(-1)^k\, x^k}{k^2} + \cdots$$

Use the Power Series Convergence Principle to confirm that the power series converges to $f[x]$ for $-1 < x < 1$.

Answer: Take $R = 1$, and look at the infinite list of individual terms

$$\left\{1, -R, \frac{R^2}{2^2}, \frac{-R^3}{3^2}, \frac{R^4}{4^2}, \ldots, \frac{(-1)^k R^k}{k^2}, \ldots\right\}$$

$$= \left\{1, -1, \frac{1}{2^2}, \frac{-1}{3^2}, \frac{1}{4^2}, \frac{-1}{5^2}, \ldots, \frac{(-1)^k}{k^2}, \ldots\right\}.$$

All the terms in this list are captured between -1 and 1. This tells you that they stay bounded, because they cannot blow up to ∞ or down to $-\infty$. The upshot: You are guaranteed that power series

$$1 - x + \frac{x^2}{2^2} - \frac{x^3}{3^2} + \cdots + \frac{(-1)^k x^k}{k^2} + \cdots$$

converges to $f[x]$ for $-1 < x < 1$.

B.4.b.ii) | Do you always have to use the Power Series Convergence Principle to determine a convergence interval for a power series?

Answer: No. If the power series comes from a differential equation as in B.3) above, you can inspect the differential equation to determine a convergence interval. If you recognize a power series as an expansion of a function $f[x]$ you know about, then you can find convergence intervals by looking at complex singularities of $f[x]$.

Case in point: If the power series

$$1 - x^2 + x^4 - x^6 + x^8 - x^{10} + \cdots + (-1)^k x^{2k} + \cdots$$

comes your way, you should recognize it as the expansion of

$$\frac{1}{1 + x^2}$$

in powers of x. Because this function has complex singularities at i and $-i$, you know that $R = 1$. Consequently the power series converges to

$$f[x] = \frac{1}{1 + x^2} \quad \text{for } -1 < x < 1.$$

You should reserve the Power Series Convergence Principle for situations you can't handle by another method.

B.4.c) | What ideas underlie the Power Series Convergence Principle?

Answer: This is going to be quite technical. Maybe you can afford not to understand it fully. After you go on in mathematics, this little discussion will not seem so obtuse.

Lesson 4.06 Power Series

The Power Series Convergence Principle says that if $f[x]$ is defined by a power series

$$a[0] + a[1]\, x + a[2]\, x^2 + a[3]\, x^3 + \cdots + a[k]\, x^k + \cdots$$

where the $a[k]$'s are given constants, then the power series converges to $f[x]$ for $-R < x < R$, provided that the infinite list of individual terms

$$\{a[0], a[1]\, R, a[2]\, R^2, a[3]\, R^3, \ldots, a[k]\, R^k, \ldots\}$$

stays bounded. Saying that this list of individual terms stays bounded, is the same as saying that you can get your hands on a number M such that M is larger than everything in the list of terms

$$\{|a[0]|, |a[1]\, R|, |a[2]\, R^2|, |a[3]\, R^3|, |a[4]\, R^4|, \ldots, |a[k]\, R^k|, \ldots\}.$$

Now take any x with $-R < x < R$ and note that $|a[k]\, x^k| \le |a[k]\, r^k|$, where r is any number with $|x| < r < R$. ($r = (R+x)/2$ will do just fine.)

Take another look:

$$|a[k]\, x^k| \le |a[k]\, r^k|$$
$$= \frac{|a[k]\, R^k|\, r^k}{|R^k|}$$
$$= |a[k]\, R^k| \left(\frac{r}{|R|}\right)^k$$
$$= |a[k]\, R^k|\, t^k$$
$$\le M\, t^k$$

where $t = r/|R|$. Make careful note of the fact that $0 < t < 1$. This ensures that the series

$$1 + t + t^2 + t^3 + t^4 + t^5 + \cdots + t^k + \cdots$$

is convergent to $1/(1-t)$. Consequently, the series

$$M + M\, t + M\, t^2 + M\, t^3 + \cdots + M\, t^k + \cdots$$

is convergent to $M/(1-t)$. Here is the situation: The term $|a[k]\, x^k|$ is under the corresponding term $M\, t^k$ of the convergent series. This tells you that the power series

$$a[0] + a[1]\, x + a[2]\, x^2 + a[3]\, x^3 + \cdots + a[k]\, x^k + \cdots$$

has no choice but to converge at least as fast as the convergent series

$$M + M\, t + M\, t^2 + M\, t^3 + M\, t^4 + \cdots + M\, t^k + \cdots$$

This tells you why

$$a[0] + a[1]\, x + a[2]\, x^2 + a[3]\, x^3 + \cdots + a[k]\, x^k + \cdots$$

is convergent provided $-R < x < R$.

The salient point of this detailed discussion is that when you go with $-R < x < R$, then the power series converges faster than a multiple of a very well-known convergent series.

Tutorials

■ T.1) Trying to plot functions defined by power series

T.1.a) If you have lots of plus and minus signs sprinkled through the terms of a power series, you can take advantage of them to get a pretty reliable plot of a function defined by a power series.

> Give a plot you believe to be trustworthy of the function $f[x]$ defined by the power series
> $$1 - \frac{x^2}{4\,(1!)} - \frac{x^4}{7\,(2!)} + \frac{x^6}{10\,(3!)} + \cdots + \frac{(-1)^k\,x^{2k}}{(3k+1)\,(k!)} + \cdots$$

Answer: Run them out so that the last term of one of them carries a plus sign and the other carries a minus sign.

In[1]:=
```
Clear[x,k,expan12,expan14]
expan12[x_] = Sum[((-1)^k) x^(2 k)/((3 k + 1) (k!)),{k,0,6}]
```
Out[1]=
$$1 - \frac{x^2}{4} + \frac{x^4}{14} - \frac{x^6}{60} + \frac{x^8}{312} - \frac{x^{10}}{1920} + \frac{x^{12}}{13680}$$

In[2]:=
```
expan14[x_] = Sum[((-1)^k) x^(2 k)/((3 k + 1) (k!)),{k,0,7}]
```
Out[2]=
$$1 - \frac{x^2}{4} + \frac{x^4}{14} - \frac{x^6}{60} + \frac{x^8}{312} - \frac{x^{10}}{1920} + \frac{x^{12}}{13680} - \frac{x^{14}}{110880}$$

The plus sign at the end of expan12[x] and the minus sign at the end of expan14[x] will help show off the differences in the two plots.

In[3]:=
```
r = 4;
Plot[{expan12[x],expan14[x]},{x,-r,r},
PlotStyle->{{Blue,Thickness[0.02]},
{Red,Thickness[0.01]}},
AxesLabel->{"x",""}];
```

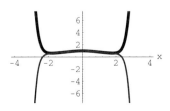

You can tell that expan12[x] is on top by looking at its highest degree term:

In[4]:=
```
expan12[x]
```

Out[4]=
$$1 - \frac{x^2}{4} + \frac{x^4}{14} - \frac{x^6}{60} + \frac{x^8}{312} - \frac{x^{10}}{1920} + \frac{x^{12}}{13680}$$

You can tell that expan14[x] is on the bottom by looking at its highest degree term:

In[5]:=
```
expan14[x]
```

Out[5]=
$$1 - \frac{x^2}{4} + \frac{x^4}{14} - \frac{x^6}{60} + \frac{x^8}{312} - \frac{x^{10}}{1920} + \frac{x^{12}}{13680} - \frac{x^{14}}{110880}$$

You go for a reasonably trustworthy plot of $f[x]$ by looking at:

In[6]:=
```
r = 4;
Plot[{expan12[x],expan14[x]},{x,-r,r},
  PlotStyle->{{Blue,Thickness[0.02]},
  {Red,Thickness[0.01]}},
  AxesLabel->{"x",""}];
```

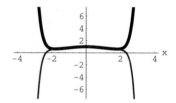

Reduce the plotting interval:

In[7]:=
```
r = 2;
Plot[{expan12[x],expan14[x]},{x,-r,r},
  PlotStyle->{{Blue,Thickness[0.02]},
  {Red,Thickness[0.01]}},
  AxesLabel->{"x",""}];
```

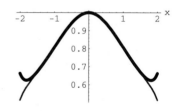

Reduce the plotting interval a bit more:

In[8]:=
```
r = 1.75;
Plot[{expan12[x],expan14[x]},{x,-r,r},
  PlotStyle->{{Blue,Thickness[0.02]},
  {Red,Thickness[0.01]}},
  AxesLabel->{"x",""}];
```

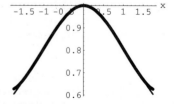

Here's what looks like a reasonably trustworthy plot of $f[x]$:

In[9]:=
```
r = 1.7;
Plot[expan14[x],{x,-r,r},
PlotStyle->{{Blue,Thickness[0.02]}},
AxesLabel->{"x","f[x]"}];
```

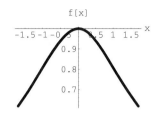

T.1.b) Sometimes even when you don't have a lot of plus and minus signs sprinkled through the terms of a power series, you can do pretty well.

> Give a plot you believe to be trustworthy of the function $f[x]$ defined by the power series
> $$1 + \frac{x}{2} + \frac{x^2}{3} + \frac{x^3}{4} + \cdots + \frac{x^k}{(k+1)} + \cdots$$

Answer: Take a look at the beginning of the expansion:

In[10]:=
```
Clear[x,k,expan10]
expan10[x_] = Sum[x^k/((k + 1)),{k,0,10}]
```

Out[10]=
$$1 + \frac{x}{2} + \frac{x^2}{3} + \frac{x^3}{4} + \frac{x^4}{5} + \frac{x^5}{6} + \frac{x^6}{7} + \frac{x^7}{8} + \frac{x^8}{9} + \frac{x^9}{10} + \frac{x^{10}}{11}$$

Even though you see no minus signs, when x is negative the signs of the terms will alternate:

In[11]:=
```
expan10[-x]
```

Out[11]=
$$1 - \frac{x}{2} + \frac{x^2}{3} - \frac{x^3}{4} + \frac{x^4}{5} - \frac{x^5}{6} + \frac{x^6}{7} - \frac{x^7}{8} + \frac{x^8}{9} - \frac{x^9}{10} + \frac{x^{10}}{11}$$

Take advantage of this by bouncing expan9[x] off expan10[x] and looking for split ends on the far left:

In[12]:=
```
r = 2; Clear[expan9]
expan9[x_] = Sum[x^k/(k + 1),{k,0,9}];
Plot[{expan9[x],expan10[x]},{x,-r,r},
PlotStyle->{{Blue},{Red}},
AxesLabel->{"x",""}];
```

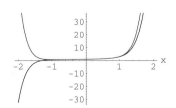

Split ends. Reduce the plotting interval:

Lesson 4.06 Power Series

In[13]:=
```
r = 1;
Plot[{expan9[x],expan10[x]},{x,-r,r},
 PlotStyle->{{Blue},{Red}},
 AxesLabel->{"x",""}];
```
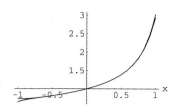

Shave the plotting interval a bit more:

In[14]:=
```
r = 0.8;
Plot[{expan9[x],expan10[x]},{x,-r,r},
 PlotStyle->{{Blue},{Red}},
 AxesLabel->{"x",""}];
```
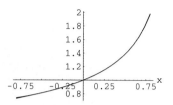

Sharing ink all the way. A reasonably trustworthy plot of $f[x]$ seems to be:

In[15]:=
```
r = 0.8;
Plot[expan10[x],{x,-r,r},
 PlotStyle->{{Blue,Thickness[0.01]}},
 AxesLabel->{"x","f[x]"}];
```
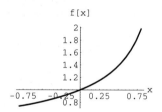

■ T.2) Functions defined by power series via differential equations

T.2.a.i) The function $y[x]$ under study here is defined to be the solution of the differential equation

$$y''[x] + x^2\, y[x] = 0 \quad \text{with } y[0] = 1 \text{ and } y'[0] = -1.$$

Mathematica cannot come up with a formula for this function:

In[16]:=
```
Clear[y,x,Derivative]
DSolve[{y''[x] + x^2 y[x] == 0,y[0] == 1,y'[0] == -1},y[x],x]
```

Out[16]=
$$\text{DSolve}[\{x^2\, y[x] + y''[x] == 0,\ y[0] == 1,\ y'[0] == -1\},\ y[x],\ x]$$

> Come up with the expansion of this function in powers of x through the x^{17} term.

Answer: Copy, paste, and edit and away you go:

In[17]:=
```
Clear[x,k,a,prelimexpan]; Clear[expan17]; m = 17;
prelimexpan[x_] = (Sum[a[k] x^k,{k,0,m}] + O[x]^(m + 1))/.{a[0]->1,a[1]->-1};
left = prelimexpan''[x] + x^2 prelimexpan[x];
right = 0; coeffeqns = LogicalExpand[left == right];
coeffsolved = Chop[Solve[coeffeqns]];
expan17[x_] = Normal[prelimexpan[x]/.coeffsolved[[1]]]
```

Out[17]=

$$1 - x - \frac{x^4}{12} + \frac{x^5}{20} + \frac{x^8}{672} - \frac{x^9}{1440} - \frac{x^{12}}{88704} + \frac{x^{13}}{224640} + \frac{x^{16}}{21288960} - \frac{x^{17}}{61102080}$$

T.2.a.ii) Use the result of part T.2.a.i) to come up with what you believe to be a trustworthy plot of the function $y[x]$ defined to be the solution of the differential equation

$$y''[x] + x^2\, y[x] = 0 \qquad \text{with } y[0] = 1 \text{ and } y'[0] = -1.$$

Check yourself with NDSolve.

Answer: Look again:

In[18]:=
```
expan17[x]
```

Out[18]=

$$1 - x - \frac{x^4}{12} + \frac{x^5}{20} + \frac{x^8}{672} - \frac{x^9}{1440} - \frac{x^{12}}{88704} + \frac{x^{13}}{224640} + \frac{x^{16}}{21288960} - \frac{x^{17}}{61102080}$$

Bounce this off the expansion through the x^{16} term:

In[19]:=
```
Clear[expan16]
expan16[x_] = Normal[Series[expan17[x],{x,0,16}]]
```

Out[19]=

$$1 - x - \frac{x^4}{12} + \frac{x^5}{20} + \frac{x^8}{672} - \frac{x^9}{1440} - \frac{x^{12}}{88704} + \frac{x^{13}}{224640} + \frac{x^{16}}{21288960}$$

And plot:

In[20]:=
```
r = 4;
Plot[{expan16[x],expan17[x]},{x,-r,r},
PlotStyle->{{Thickness[0.02],Blue},
{Thickness[0.01],Red}},
AxesLabel->{"x",""}];
```

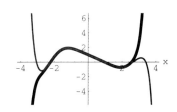

208 Lesson 4.06 Power Series

Cut back the plotting interval:

In[21]:=
```
r = 2.5;
Plot[{expan16[x],expan17[x]},{x,-r,r},
PlotStyle->{{Thickness[0.02],Blue},
{Thickness[0.01],Red}},
AxesLabel->{"x",""}];
```

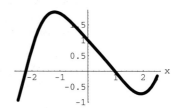

Here's a trustworthy plot:

In[22]:=
```
r = 2.4;
trusty = Plot[expan17[x],{x,-r,r},
PlotStyle->{{Thickness[0.02],Blue}},
AxesLabel->{"x","y[x]"}];
```

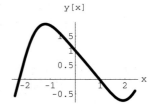

Check it with the plot from NDSolve:

In[23]:=
```
a = -r; c = r;
Clear[solution,x,y,fakey,Derivative];
solution = NDSolve[{y''[x] + x^2 y[x] == 0,
y[0] == 1,y'[0] == -1},y[x],{x,a,c}];
fakey[x_] = y[x]/.solution[[1]];
Mathematicaplot = Plot[fakey[x],{x,a,c},
PlotStyle->{{Red,Thickness[0.01]}},
DisplayFunction->Identity];
Show[trusty,Mathematicaplot,
AxesLabel->{"x","y[x]"}];
```

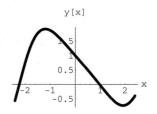

It doesn't get much better than that.

T.2.a.iii) Here is *Mathematica*'s plot of the solution of the differential equation
$$y''[x] + x^2 y[x] = 0 \quad \text{with } y[0] = 1 \text{ and } y'[0] = -1 \text{ for } -6 \leq x \leq 6:$$

In[24]:=
```
r = 6; a = -r; c = r;
Clear[solution,x,y,fakey,Derivative]
solution = NDSolve[{y''[x] + x^2 y[x] == 0,
y[0] == 1,y'[0] == -1},y[x],{x,a,c}];
fakey[x_] = y[x]/.solution[[1]];
Plot[fakey[x],{x,a,c},
PlotStyle->{{Red,Thickness[0.01]}},
AxesLabel->{"x","y[x]"}];
```

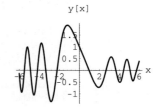

> Is it possible to reproduce this plot by plotting partial expansions?

Answer: In theory, the answer is yes.

Reason: Theory, as found in B.3), says that the convergence interval of the expansion of the function $y[x]$ defined by the differential equation $y''[x] + x^2\, y[x] = 0$ with $y[0] = 1$ and $y'[0] = -1$ is $-\infty < x < \infty$.

In practice, the answer is no.

Reason: To even make an attempt at the plot on this long interval, you would need a big, fast computer that could handle all the calculations involved in plotting really long partial expansions. It's one thing to plot the expansion through the x^{13} term. It's quite another thing to plot the expansion through the x^{50} term.

■ T.3) Using the Power Series Convergence Principle: The Ratio Test

Given a power series

$$a[0] + a[1]\,x + a[2]\,x^2 + \cdots + a[k]\,x^k + \cdots,$$

you can attempt to find some of its convergence intervals by using the Power Series Convergence Principle or by using the Ratio Test. The Ratio Test says:

If, for some positive number R, you can ascertain that

$$\left|\frac{a[k+1]\,R^{k+1}}{a[k]\,R^k}\right| \leq 1$$

for all large k's, then

$$a[0] + a[1]\,x + a[2]\,x^2 + \cdots + a[k]\,x^k + \cdots$$

converges for $-R < x < R$.

T.3.a) > Find convergence intervals of the following power series. Use the Power Series Convergence Principle directly, use the Ratio Test, or use any other method you like.

T.3.a.i) > $1 - x + x^2 - x^3 + x^4 - x^5 + \cdots + (-1)^n x^n + \cdots$

Answer:

$\rightarrow$ By the Power Series Convergence Principle:

The power series is
$$1 - x + x^2 - x^3 + x^4 - x^5 + \cdots + (-1)^n x^n + \cdots.$$
For $R = 1$, the list of terms
$$\{1, -R, R, -R^3, R^4, \ldots, (-1)^n R^n, \ldots\} = \{1, -1, 1, -1, 1, -1, \ldots, (-1)^n, \ldots\}$$
stays bounded. This guarantees that the power series converges for $-1 < x < 1$.

→ By the Ratio Test:

The power series is
$$1 - x + x^2 - x^3 + x^4 - x^5 + \cdots + (-1)^n x^n + \cdots.$$
Note that $|a[n]| = 1$ for all n's. For a positive number R, look at
$$\frac{|a[n+1] R^{n+1}|}{|a[n] R^n|} = \frac{R^{n+1}}{R^n} = R.$$
These ratios are all ≤ 1 for $R = 1$. This guarantees that the power series converges for $-1 < x < 1$.

→ The easiest way:

The power series is
$$1 - x + x^2 - x^3 + x^4 - x^5 + \cdots + (-1)^n x^n + \cdots.$$
You already know that this is the expansion of $1/(1+x)$ in powers of x and that it converges for $-1 < x < 1$.

Each of these three answers is complete. Explaining the truth requires only one good argument. Explaining lies usually requires many long, involved points of view.

T.3.a.ii)
$$\frac{x^2}{2 \times 3^2} + \frac{x^3}{3 \times 3^3} + \cdots + \frac{x^k}{k \times 3^k} + \cdots$$

Answer: Go with the Ratio Test: For $k > 2$, read off:

In[25]:=
```
Clear[R,k,a]; a[k_] = 1/(k 3^k)
```

Out[25]=
$$\frac{1}{3^k k}$$

For a positive number R, look at:

In[26]:=
```
ratio = Simplify[a[k+1] R^(k + 1)/(a[k] R^k)]
```

Out[26]=
$$\frac{kR}{3(1+k)}$$

Look at the ratios for $R = 3$:

In[27]:=
```
ratio/.R->3
```

Out[27]=
$$\frac{k}{1+k}$$

Evidently, these ratios are ≤ 1 for $R = 3$. This guarantees that the power series converges for $-3 < x < 3$.

Just for kicks, here's a second argument using the Power Series Convergence Principle:

The power series is:

$$\frac{x^2}{2 \times 3^2} + \frac{x^3}{3 \times 3^3} + \cdots + \frac{x^k}{k \times 3^k} + \cdots$$

For $R = 3$, the list of terms

$$\left\{ \frac{R^2}{2 \times 3^2}, \frac{R^3}{3 \times 3^3}, \frac{R^4}{4 \times 3^4}, \ldots, \frac{R^k}{n \times 3^k}, \ldots \right\} = \left\{ \frac{3^2}{2 \times 3^2}, \frac{3^3}{3 \times 3^3}, \frac{3^4}{4 \times 3^4}, \ldots, \frac{3^k}{k \times 3^k}, \ldots \right\}$$

$$= \left\{ \frac{1}{2}, \frac{1}{3}, \frac{1}{4}, \frac{1}{5}, \ldots, \frac{1}{k}, \ldots \right\}$$

stays bounded. This guarantees that the power series converges for $-3 < x < 3$.

T.3.a.iii)
$$1 + x + \frac{x^2}{2^2} + \frac{x^3}{3^3} + \frac{x^4}{4^4} + \cdots + \frac{x^k}{k^k} + \cdots$$

Answer: The Power Series Convergence Principle works like a charm. For any positive number $x = R$, look at the list of terms

$$\left\{ 1, R, \frac{R^2}{2^2}, \frac{R^3}{3^3}, \frac{R^4}{4^4}, \ldots, \frac{R^k}{k^k}, \ldots \right\}.$$

Notice that after the point at which n becomes larger than R, the terms get smaller and smaller. Take a look at the case $R = 9$:

In[28]:=
```
Clear[k]
ListPlot[Table[(9^n)/n^n,{n,1,25}],
PlotStyle->{PointSize[0.02],Red}];
```

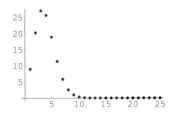

Take a look at the case $R = 30$:

In[29]:=
```
Clear[k]
ListPlot[Table[(30^n)/n^n,{n,1,60}],
 PlotStyle->{PointSize[0.02],Red}];
```

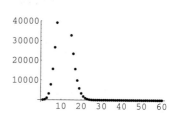

No matter what positive R you take, the list of terms

$$\left\{1, R, \frac{R^2}{2^2}, \frac{R^3}{3^3}, \frac{R^4}{4^4}, \ldots, \frac{R^k}{k^k}, \ldots\right\}$$

cannot blow up to ∞ or down to $-\infty$. The Power Series Convergence Principle steps in to guarantee that no matter what positive R you go with, this power series converges for $-R < x < R$.

The upshot: This power series converges for $-\infty < x < \infty$.

T.3.a.iv) $\quad 1 - 2^2\, x + 3^2\, x^2 - 4^2\, x^3 + \cdots + (-1)^k\, (k+1)^2\, x^k + \cdots$

Answer:

$\rightarrow$ By the Power Series Convergence Principle:

Take any positive R and look at the list of terms:
$$\{1, -2^2\, R, 3^2\, R^2, -4^2\, R^3, 5^2\, R^4, \ldots (-1)^k\, (k+1)^2\, R^k, \ldots\}.$$
Put $R = e^t$ to get
$$\{1, -2^2\, e^t, 3^2\, e^{2t}, -4^2\, e^{3t}, 5^2\, e^{4t}, \ldots (-1)^k\, (k+1)^2\, e^{kt}, \ldots\}.$$
If $t < 0$, then
$$\lim_{k \to \infty} (-1)^k\, (k+1)^2\, e^{kt} = 0$$
because exponential growth dominates power growth. So, if $t < 0$, the list
$$\{1, -2^2\, e^t, 3^2\, e^{2t}, -4^2\, e^{3t}, 5^2\, e^{4t}, \ldots, (-1)^k\, (k+1)^2\, e^{kt}, \ldots\}$$
cannot blow up to $+\infty$ or down to $-\infty$.

The result: The power series converges for $-e^t < x < e^t$ as long as $t < 0$. But as t closes in on 0 through negative numbers, e^t closes in on 1. The upshot: The power series converges for $-1 < x < 1$.

$\rightarrow$ By the Ratio Test:

In[30]:=
```
Clear[a,k,R]; a[k_] = (-1)^k (k + 1)^2
```
Out[30]=
$$(-1)^k\, (1 + k)^2$$

In[31]:=
```
firstlook = Cancel[a[k + 1] R^(k + 1)/(a[k] R^k)]
```
Out[31]=
$$-\left(\frac{(2+k)^2 R}{(1+k)^2}\right)$$

In[32]:=
```
secondlook = Together[ExpandAll[firstlook]]
```
Out[32]=
$$\frac{-4R - 4kR - k^2 R}{1 + 2k + k^2}$$

So,
$$\lim_{k \to \infty} \left| \frac{a[k+1]\, R^{k+1}}{a[k]\, R^k} \right| = R.$$

As a result, if R is any number with $0 \le R < 1$, then
$$\left| \frac{a[k+1]\, R^{k+1}}{a[n]\, R^n} \right| < 1$$
for all large n's.

The Ratio Test tells you that if R is any number with $0 \le R < 1$, then this power series converges for $-R < x < R$. In short, this power series converges for $-1 < x < 1$.

T.3.a.v)
$$1 + 2x + \frac{2^2 x^2}{2!} + \frac{2^3 x^3}{3!} + \cdots + \frac{2^k x^k}{k!} + \cdots$$

Answer: This is a soft pitch. Recall that
$$1 + x + \frac{x^2}{2!} + \frac{x^3}{3!} + \frac{x^4}{4!} + \cdots + \frac{x^k}{k!} + \cdots$$
is the expansion of e^x in powers of x. It converges to e^x for $-\infty < x < \infty$. Replacing x by $2x$ above shows that the power series in question is the expansion of e^{2x} in powers of x. It converges to e^{2x} for $-\infty < x < \infty$. Consequently, the power series
$$1 + 2x + \frac{2^2 x^2}{2!} + \frac{2^3 x^3}{3!} + \cdots + \frac{2^k x^k}{k!} + \cdots$$
converges for $-\infty < x < \infty$.

T.3.b) How do you use the Power Series Convergence Principle to see why the Ratio Test works?

Answer: The Ratio Test says:

If for some positive number R, you can ascertain that
$$\left|\frac{a[k+1]\,R^{k+1}}{a[k]\,R^k}\right| \leq 1$$
for all large k's, then $a[0] + a[1]\,x + a[2]\,x^2 + \cdots + a[k]\,x^k + \cdots$ converges for $-R < x < R$.

Saying that
$$\left|\frac{a[k+1]\,R^{k+1}}{a[k]\,R^k}\right| \leq 1$$
for all large k's is the same as saying that
$$|a[k+1]\,R^{k+1}| \leq |a[k]\,R^k|$$
for all large k's. This tells you that the infinite list of individual terms
$$\{a[1]\,R, a[2]\,R^2, a[3]\,R^3, \ldots, a[k]\,R^k, \ldots\}$$
cannot blow up to ∞ or down to $-\infty$, because after some point in the list, the absolute values of the terms cannot increase. The Basic Power Series Convergence Principle steps in to guarantee that
$$a[0] + a[1]\,x + a[2]\,x^2 + \cdots + a[k]\,x^k + \cdots$$
converges for $-R < x < R$. That's all there is to it.

■ T.4) Infinite sums of numbers

T.4.a) Explain why the following infinite sum of numbers is convergent. Try to estimate the sum.
$$-2 + \frac{7}{8} - \frac{5}{24} + \cdots + \frac{(-1)^k\,(3\,k+1)}{2^k\,k!} + \cdots$$

Answer: Look at the power series
$$-2\,x + \frac{7}{8}\,x^2 - \frac{5}{24}\,x^3 + \cdots + \frac{(-1)^k\,(3\,k+1)}{2^n\,k!}\,x^k + \cdots$$
If you can determine that this power series converges on an interval including $x = 1$, then you'll have explained why
$$-2 + \frac{7}{8} - \frac{5}{24} + \cdots + \frac{(-1)^k\,(3\,k+1)}{2^k\,k!} + \cdots$$
converges. To this end, try to determine some convergence intervals of
$$-2\,x + \frac{7}{8}\,x^2 - \frac{5}{24}\,x^3 + \cdots + \frac{(-1)^k\,(3\,k+1)}{2^k\,k!}\,x^k + \cdots$$

The Ratio Test is a good way to do this one.

In[33]:=
```
Clear[a,k,R]; a[k_] = (-1)^k (3 k + 1)/(2^k k!)
```
Out[33]=
$$\frac{(-1)^k (1 + 3k)}{2^k \, k!}$$

Take a positive R and look at $a[k+1]\,R^{k+1} / \left(a[k]\,R^k\right)$:

In[34]:=
```
Simplify[a[k + 1] R^(k+1)/(a[k] R^k)]
```
Out[34]=
$$\frac{-((4 + 3k) \, R \, k!)}{2 \, (1 + 3k) \, (1 + k)!}$$

So, $|a[k+1]\,R^{k+1}/(a[k]\,R^k)|$ is

$$\frac{R(4+3k)}{2(1+3k)(k+1)}$$

To cancel the fraction remember that $(k+1)! = (k+1)\,(k!)$; so

$$\frac{k!}{(k+1)!} = \frac{1}{k+1}.$$

In[35]:=
```
firstshot = ExpandAll[R (4 + 3 k)/(2 (1 + 3 k) (k + 1))]
```
Out[35]=
$$\frac{4R}{2 + 8k + 6k^2} + \frac{3kR}{2 + 8k + 6k^2}$$

In[36]:=
```
bettershot = Together[firstshot]
```
Out[36]=
$$\frac{4R + 3kR}{2(1 + 4k + 3k^2)}$$

So, no matter what R is,

$$\lim_{k \to \infty} \left| \frac{a[k+1]\,R^{k+1}}{a[k]\,R^k} \right| = \lim_{k \to \infty} \frac{3kR}{6k^2} = 0.$$

Consequently, no matter what R is,

$$\left| \frac{a[k+1]\,R^{k+1}}{a[k]\,R^k} \right| \leq 1 \qquad \text{for all large } k\text{'s.}$$

Lesson 4.06 Power Series

This tells you that the power series

$$-2x + \frac{7}{8}x^2 - \frac{5}{24}x^3 + \cdots + \frac{(-1)^n (3n+1)}{2^n n!} x^n + \cdots$$

converges for $-\infty < x < \infty$. Since $x = 1$ satisfies $-\infty < x < \infty$, the infinite sum

$$-2 + \frac{7}{8} - \frac{5}{24} + \cdots + \frac{(-1)^k (3k+1)}{2^k k!} + \cdots$$

is convergent (to $f[1]$, where $f[x]$ is the function defined by the power series). Watch it converge:

In[37]:=
N[Sum[a[k],{k,1,1}],12]

Out[37]=
-2.

In[38]:=
N[Sum[a[k],{k,1,7}],12]

Out[38]=
-1.30326760913

In[39]:=
N[Sum[a[k],{k,1,10}],12]

Out[39]=
-1.30326532946

This one seems to be converging like the proverbial bat out of the bad place. No wonder; look at all the power in those denominators!

In[40]:=
N[Sum[a[k],{k,1,15}],18]

Out[40]=
-1.30326532985631675

In[41]:=
N[Sum[a[k],{k,1,20}],30]

Out[41]=
-1.30326532985631671180189918407

Settling down beautifully.

In[42]:=
N[Sum[a[k],{k,1,30}],30]

Out[42]=
-1.30326532985631671180189918407

A very confident educated guess is that

$$-2 + \frac{7}{8} - \frac{5}{24} + \cdots + \frac{(-1)^k (3k+1)}{2^k k!} + \cdots = -1.303265329856$$

to 12 accurate decimals.

T.4.b) Explain why the following infinite sum of numbers is convergent. Try to estimate the sum.
$$1 + \frac{1}{10} + \frac{1}{125} + \frac{1}{1250} + \frac{1}{10625} \cdots + \frac{1}{5^k(k^2+1)} + \cdots$$

Answer: Look at the power series
$$1 + \frac{x}{10} + \frac{x^2}{125} + \frac{x^3}{1250} + \frac{x^4}{10625} \cdots + \frac{x^k}{5^k(k^2+1)} + \cdots$$

If you can determine that this power series converges on an interval including $x = 1$, then you'll have explained why
$$1 + \frac{1}{10} + \frac{1}{125} + \frac{1}{1250} + \frac{1}{10625} \cdots + \frac{1}{5^k(k^2+1)} + \cdots$$
converges. To this end, try to determine some convergence intervals of
$$1 + \frac{x}{10} + \frac{x^2}{125} + \frac{x^3}{1250} + \frac{x^4}{10625} \cdots + \frac{x^k}{5^k(k^2+1)} + \cdots$$

The Power Series Convergence Principle is a good way to do this one. Take a positive R. The list of terms is:
$$\left\{ 1, \frac{R}{10}, \frac{R^2}{125}, \frac{R^3}{1250}, \frac{R^4}{10625}, \ldots, \frac{R^k}{5^k(k^2+1)}, \ldots \right\}.$$

When you take $R = 5$, the list is
$$\left\{ 1, \frac{5}{10}, \frac{25}{125}, \frac{125}{1250}, \frac{625}{10625}, \ldots, \frac{5^k}{5^k(k^2+1)}, \ldots \right\}$$
$$= \left\{ 1, \frac{5}{10}, \frac{25}{125}, \frac{125}{1250}, \frac{625}{1062}, \ldots, \frac{1}{k^2+1}, \ldots \right\}.$$

All the numbers in this list are captured between 0 and 1. The upshot:
$$1 + \frac{x}{10} + \frac{x^2}{125} + \frac{x^3}{1250} + \frac{x^4}{10625} \cdots + \frac{x^k}{5^k(k^2+1)} + \cdots$$
converges for $-5 < x < 5$. Since $-5 < 1 < 5$, this tells you that the infinite sum
$$1 + \frac{1}{10} + \frac{1}{125} + \frac{1}{1250} + \frac{1}{10625} \cdots + \frac{1}{5^k(k^2+1)} + \cdots$$
converges (to $f[1]$, where $f[x]$ is the function defined by the power series). Watch it converge:

In[43]:=
```
Clear[a,k]; a[k_] = 1/(5^k (k^2 + 1))
```

Out[43]=

$$\frac{1}{5^k (1 + k^2)}$$

In[44]:=
`N[Sum[a[k],{k,1,3}],12]`

Out[44]=
0.1088

In[45]:=
`N[Sum[a[k],{k,1,20}],20]`

Out[45]=
0.10890845791341907116

In[46]:=
`N[Sum[a[k],{k,1,30}],30]`

Out[46]=
0.108908457913419076966620411227

In[47]:=
`N[Sum[a[k],{k,1,50}],30]`

Out[47]=
0.108908457913419076966620686072

A very confident educated guess is that the infinite sum

$$1 + \frac{1}{10} + \frac{1}{125} + \frac{1}{1250} + \frac{1}{10625} \cdots + \frac{1}{5^k (k^2 + 1)} + \cdots$$

converges to 0.1089084579 to 10 accurate decimals.

T.4.c) Try to explain why the infinite sum

$$1 + \frac{1}{2} + \frac{1}{3} + \frac{1}{4} + \cdots + \frac{1}{k} + \cdots$$

is not convergent. Fancy folks call this the harmonic series.

Answer: Note:

$$x + \frac{x^2}{2} + \frac{x^3}{3} + \frac{x^4}{4} + \cdots + \frac{x^n}{n} + \cdots = \int_0^x \left(1 + t + t^2 + t^3 + \cdots + t^{n-1} + \cdots\right) dt$$

$$= \int_0^x \frac{1}{1-t} \, dt \qquad \text{provided } 0 \leq x < 1.$$

$$= -\log[1-x]$$

This reveals that, for $0 \leq x < 1$,

$$x + \frac{x^2}{2} + \frac{x^3}{3} + \frac{x^4}{4} + \cdots + \frac{x^n}{n} + \cdots = -\log[1-x].$$

Now
$$1 + \frac{1}{2} + \frac{1}{3} + \frac{1}{4} + \cdots + \frac{1}{k} + \cdots = \lim_{x \to 1} \left(x + \frac{x^2}{2} + \frac{x^3}{3} + \cdots + \frac{x^n}{n} + \cdots \right)$$
$$= \lim_{x \to 1} \left(-\log[1-x] \right)$$
$$= \infty.$$

This tells you that the infinite sum
$$1 + \frac{1}{2} + \frac{1}{3} + \frac{1}{4} + \cdots + \frac{1}{k} + \cdots$$
blows up and cannot converge. See what happens when you take some partial sums:

In[48]:=
    `Clear[k]; N[Sum[1/k,{k,1,2}]]`

Out[48]=
    1.5

In[49]:=
    `N[Sum[1/k,{k,1,4}]]`

Out[49]=
    2.08333

In[50]:=
    `N[Sum[1/k,{k,1,2^3}]]`

Out[50]=
    2.71786

In[51]:=
    `N[Sum[1/k,{k,1,2^4}]]`

Out[51]=
    3.38073

In[52]:=
    `N[Sum[1/k,{k,1,2^6}]]`

Out[52]=
    4.74389

In[53]:=
    `N[Sum[1/k,{k,1,2^8}]]`

Out[53]=
    6.12434

In[54]:=
    `N[Sum[1/k,{k,1,2^10}]]`

Out[54]=
    7.50917567227813

The partial sums are not settling down. They are going up just a little faster than $\log[x]$ goes up:

In[55]:=
{N[Sum[1/k,{k,1,2^3}]],N[Log[2^3]]}

Out[55]=
{2.71786, 2.07944}

In[56]:=
{N[Sum[1/k,{k,1,2^4}]],N[Log[2^4]]}

Out[56]=
{3.38073, 2.77259}

In[57]:=
{N[Sum[1/k,{k,1,2^6}]],N[Log[2^6]]}

Out[57]=
{4.74389, 4.15888}

In[58]:=
{N[Sum[1/k,{k,1,2^8}]],N[Log[2^8]]}

Out[58]=
{6.12434, 5.54518}

In[59]:=
{N[Sum[1/k,{k,1,2^10}]],N[Log[2^10]]}

Out[59]=
{7.50917567227813, 6.93147}

You might be able to see why these sums behave this way. Think about it.

If you want to see another intriguing reason why the infinite sum

$$1 + \frac{1}{2} + \frac{1}{3} + \frac{1}{4} + \cdots + \frac{1}{k} + \cdots$$

does not converge, then look at this. Some folks love this argument.

Very tricky way of seeing why $1 + \frac{1}{2} + \frac{1}{3} + \frac{1}{4} + \cdots + \frac{1}{k} + \cdots$ does not converge

Assume it does converge and then try for a contradiction. To this end suppose it converges to a definite value s. Thus

$$s = 1 + \frac{1}{2} + \frac{1}{3} + \frac{1}{4} + \cdots$$

Divide through by 2 to get

$$\frac{s}{2} = \frac{1}{2} + \frac{1}{4} + \frac{1}{6} + \frac{1}{8} + \frac{1}{10} + \cdots.$$

Subtract the expression for $s/2$ from the expression for s. This gives

$$\frac{s}{2} = s - \frac{s}{2} = 1 + \frac{1}{3} + \frac{1}{5} + \frac{1}{7} + \cdots$$

This gives you two expressions for $s/2$:

$$\frac{s}{2} = 1 + \frac{1}{3} + \frac{1}{5} + \frac{1}{7} + \cdots$$

$$\frac{s}{2} = \frac{1}{2} + \frac{1}{4} + \frac{1}{6} + \frac{1}{8} + \cdots$$

Note that each term in the first expression for $s/2$ is strictly greater than the corresponding term in the second expression. The inescapable conclusion is that

$$\frac{s}{2} > \frac{s}{2}.$$

But there is no number s with this property. This means that there is no number s with

$$s = 1 + \frac{1}{2} + \frac{1}{3} + \frac{1}{4} + \cdots$$

This is another reason why the harmonic series

$$1 + \frac{1}{2} + \frac{1}{3} + \frac{1}{4} + \cdots$$

cannot converge.

Give It a Try

Experience with the starred ($\star$) problems will be especially beneficial for understanding later lessons.

■ G.1) Trying to plot functions defined by power series*

To see a whole zoo of functions defined by power series, look at the book, *Handbook of Mathematical Functions with Formulas, Graphs, and Mathematical Tables*, Milton Abramowitz and Irene A. Stegun editors, National Bureau of Standards, Wiley-Interscience, New York, 1972.

G.1.a) Give a plot you believe to be trustworthy of the function $f[x]$ defined by the power series

$$1 + \frac{x}{2!} + \frac{x^2}{4!} + \frac{x^3}{6!} + \cdots + \frac{x^k}{(2k)!} + \cdots$$

G.1.b) The function called the Struve function $f[x]$ is defined by the power series

$$x - \frac{x^3}{1^2\, 3^2} + \frac{x^5}{1^2\, 3^2\, 5^2} - \frac{x^7}{1^2\, 3^2\, 5^2\, 7^2} + \cdots$$

> Give a plot you believe to be trustworthy of the Struve function for $-3 \leq x \leq 3$.

■ G.2) Taylor's formula and power series*

G.2.a.i) A function $f[x]$ is defined by the power series

$$\frac{x}{2} + \frac{x^2}{12} + \frac{x^3}{240} + \frac{x^4}{10080} \cdots + \frac{k\,x^k}{(2\,k)!} + \cdots$$

In other words,

$$\frac{x}{2} + \frac{x^2}{12} + \frac{x^3}{240} + \frac{x^4}{10080} \cdots + \frac{k\,x^k}{(2k)!} + \cdots$$

is the expansion of $f[x]$ in powers of x.

> What is the exact value of $f[0]$?

G.2.a.ii) Remember Taylor's formula:

In[1]:=
```
Clear[f,x]; Normal[Series[f[x],{x,0,7}]]
```
Out[1]=

$$f[0] + x\,f'[0] + \frac{x^2\,f''[0]}{2} + \frac{x^3\,f^{(3)}[0]}{6} + \frac{x^4\,f^{(4)}[0]}{24} + \frac{x^5\,f^{(5)}[0]}{120} + \frac{x^6\,f^{(6)}[0]}{720} + \frac{x^7\,f^{(7)}[0]}{5040}$$

> Use Taylor's formula to calculate the exact values of higher derivatives
>
> $$f'[0], f''[0], f'''[0], \ldots, f^{(6)}[0], \text{ and } f^{(7)}[0]$$
>
> of the function $f[x]$ defined by the power series
>
> $$\frac{x}{2} + \frac{x^2}{12} + \frac{x^3}{240} + \frac{x^4}{10080} \cdots + \frac{k\,x^k}{(2\,k)!} + \cdots$$

■ G.3) A differential equations medley

G.3.a) Not even mighty *Mathematica* can come up with a formula for the solution of the little differential equation

$$y'[x] = y[x]^2 - 1 \quad \text{with } y[0] = 0.9:$$

In[2]:=
```
Clear[x,y,Derivative]
DSolve[{y'[x] == y[x]^2 - 1,y[0] == 0.9},y[x],x]
```

Out[2]=
```
DSolve[{y'[x] == -1 + y[x]^2, y[0] == 0.9}, y[x], x]
```

> Use NDsolve to give a plot of the solution for $-2 \leq x \leq 2$.
>
> Find the expansion in powers of x of the solution in powers of x through the x^{16} term.
>
> Show the plot of the expansion together with the plot you got from NDSolve and comment on the similarities and differences, and the reasons for them.

G.3.b.i) *Mathematica* can't come up with a formula for the solution of the pendulum differential equation

$$y''[t] = -\sin[y[t]] \quad \text{with } y[0] = 0 \text{ and } y'[0] = 1:$$

In[3]:=
```
Clear[t,y,Derivative]
DSolve[{y''[t] == -Sin[y[t]],y[0] == 0,y'[0] == 1},y[t],t]
DSolve::dnim:
    Built-in procedures cannot solve this differential equation.
```

Out[3]=
```
DSolve[{y''[t] == -Sin[y[t]], y[0] == 0, y'[0] == 1}, y[t], t]
```

This is with good reason. No live or dead person has been able to come up with a clean formula for the solution of this beauty.

> Find the expansion of the solution, in powers of t, through the t^{11} term.
>
> Use what you get to give a trustworthy plot of $y[t]$, and check yourself with NDSolve.

G.3.b.ii)
> Use NDSolve to get a plot of the solution of the pendulum differential equation
>
> $$y''[t] = -\sin[y[t]] \quad \text{with } y[0] = 0 \text{ and } y'[0] = 1 \text{ for } -20 \leq t \leq 20.$$
>
> Why wouldn't you want to use power series to come up with this plot?

G.3.c.i)
> Use power series to find the exact solution of the differential equation
>
> $$y''[x] = x^2 - 6x + 5 \quad \text{with } y[0] = 1 \text{ and } y'[0] = -3.$$

G.3.c.ii) Find the exact solution of the differential equation
$$y''[x] = x^2 - 6x + 5 \quad \text{with } y[0] = 1 \text{ and } y'[0] = -3$$
by integrating twice.

■ G.4) Convergence intervals for power series coming from differential equations

For the following differential equations, give guaranteed convergence intervals for the expansion of the solutions in powers of x.

G.4.a) $y''[x] - y[x] = \cos[x]$ with $y[0] = 3$ and $y'[0] = -1$.

G.4.b) $(1-x)\,y''[x] - \cos[2x]\,y[x] = \sin[x^2]$ with $y[0] = 1$ and $y'[0] = -0.5$.

G.4.c) $y''[x] + e^{-x} y'[x] - 4x^3 y[x] = \dfrac{1}{1+2x^2}$ with $y[0] = -3$ and $y'[0] = -1$.

■ G.5) Guaranteed convergence intervals for power series

Find convergence intervals for the following power series. Use the Power Series Convergence Principle directly, use the Ratio Test, or use any other method you like.

G.5.a) $1 - x + \dfrac{x^2}{2} - \dfrac{x^3}{3} + \dfrac{x^4}{4} - \cdots + \dfrac{(-1)^n x^n}{n} + \cdots$

G.5.b) $1 + \dfrac{x}{e} + \dfrac{x^2}{e^2} + \dfrac{x^3}{e^3} + \cdots + \dfrac{x^k}{e^k} + \cdots$

G.5.c) $1 - \dfrac{x}{2!} + \dfrac{x^2}{4!} - \dfrac{x^3}{6!} + \cdots + \dfrac{x^k}{(2k)!} + \cdots$

G.5.d)
$$1 - 2^3 x + 3^3 x^2 - 4^3 x^3 + \cdots + (-1)^k (k+1)^3 x^k + \cdots$$

G.5.e)
$$1 - \frac{2^2 x^2}{2!} + \frac{2^4 x^4}{4!} - \cdots + \frac{(-1)^{k+1} (2x)^{2k}}{(2k)!} + \cdots$$

■ G.6) The simple Airy function

George Airy held the historic Lucasian Professorship of Applied Mathematics at Cambridge University in England in the mid 1800's. This position was held by Isaac Newton in the past and by Stephen Hawking today.

G.6.a.i) The simple Airy function is defined to be the solution of the differential equation

$$y''[x] - x\, y[x] = 0 \quad \text{with } y[0] = 1 \text{ and } y'[0] = 1.$$

> Use the method of undetermined coefficients to come up with the expansion of simple Airy function through the x^{16} term.

G.6.a.ii)
> Use the expansion of the simple Airy function to give what you believe to be a trustworthy plot of the simple Airy function.
>
> Check yourself with NDSolve.

G.6.a.iii)
> What is the guaranteed interval of convergence of the power series that defines the simple Airy function? Is the answer of much practical importance?

G.6.a.iv) As above, the simple Airy function is defined to be the solution of the differential equation $y''[x] - x\, y[x] = 0$ with $y[0] = 1$ and $y'[0] = 1$.

> Try to come up with a trustworthy plot of the derivative of the simple Airy function.

G.6.a.v) Here is a plot of the simple Airy function for $-5 \leq x \leq 5$:

Lesson 4.06 Power Series

In[4]:=
```
a = -5; c = 5;
Clear[solution,x,y,fakey,Derivative]
solution = NDSolve[{y''[x] == x y[x],
  y[0] == 1,y'[0] == 1},y[x],{x,a,c}];
fakey[x_] = y[x]/.solution[[1]];
airyplot = Plot[fakey[x],{x,a,c},
  PlotStyle->{{Red,Thickness[0.01]}},
  AxesLabel->{"x","y[x]"},PlotRange->All];
```

Looks like exponential growth or better on the right. Take a look on the left:

In[5]:=
```
a = -10; c = 2;
Clear[solution,x,y,fakey,Derivative]
solution = NDSolve[{y''[x] == x y[x],
  y[0] == 1,y'[0] == 1},y[x],{x,a,c}];
fakey[x_] = y[x]/.solution[[1]];
airyplot = Plot[fakey[x],{x,a,c},
  PlotStyle->{{Red,Thickness[0.01]}},
  AxesLabel->{"x","y[x]"},PlotRange->All];
```

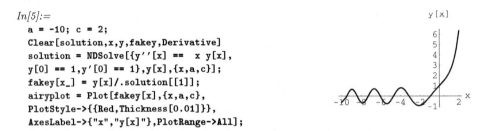

Oscillatory on the left, but incredible growth on the right.

Got any idea why the simple Airy function plots out this way?

■ G.7) Infinite sums of numbers

G.7.a) Explain why the following infinite sum of numbers is convergent. Try to estimate the sum.

$$1 + 4 + \frac{9}{2} + \frac{8}{3} + \frac{25}{24} + \frac{3}{10} + \frac{49}{720} \cdots + \frac{k^3}{k!} + \cdots$$

G.7.b) Explain why the following infinite sum of numbers is convergent. Try to estimate the sum.

$$1 - \frac{1}{6} + \frac{1}{81} - \frac{1}{756} + \frac{1}{5265} + \cdots + \frac{(-1)^k}{3^k (k^3 + 1)} + \cdots$$

G.7.c) Try to explain why the infinite sum

$$\frac{1}{2} + \frac{1}{4} + \frac{1}{6} + \frac{1}{8} + \frac{1}{2k} + \cdots$$

is not convergent.

■ G.8) A couple of bricks short of a load★

Two civil engineering professors bump into each other in the *Mathematica* lab. The younger professor says to the older: "I've got a problem in which I can get the beginning of the expansion of a function in powers of x, but I can't get the whole thing. In fact all I can get is

$$15.602 + 1.217\,x - 0.516\,x^2 - 0.171\,x^3 + 0.0416\,x^4 + 0.0107\,x^5$$
$$-0.0014\,x^6 - 0.00029\,x^7 + 0.0000248\,x^8 + 2.755\,(10^{-6})\,x^9 - 9.568\,(10^{-7})\,x^{10}.$$

I want to get as accurate a plot of the function $f[x]$ defined by the power series as I can. What should I do?" With a smile of confidence resulting from years of experience, the older professor says: "Back in the good old days before *Mathematica*, we were happy to get our expansions through the x^4 or x^5 terms. You've got plenty of information. Just plot what you have." Taking his advice, the younger professor plots:

In[6]:=
```
Clear[expan10,x]
expan10[x_] = 15.602 + 1.217 x - 0.516 x^2 -
  0.171 x^3 +0.0416 x^4 + 0.0107 x^5 -
  0.0014 x^6 - 0.00029 x^7 + 0.0000248 x^8 +
  2.755 (10^-6) x^9 - 9.568(10^-7) x^10;
Plot[expan10[x],{x,-5.5,5.5},
  PlotStyle->{{Firebrick,Thickness[0.01]}},
  PlotRange->All,AxesLabel->{"x","f[x]"}];
```

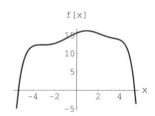

Looking over the younger professor's shoulder, the older professor says, "There you go," and draws himself a cup of coffee.

What is your reaction to this conversation? If you are the person who is giving advice to the younger professor, what advice do you give?

■ G.9) e, sin[x], and cos[x] from an advanced viewpoint★

By now most of you are becoming tired of reading that

$$e^x = 1 + x + \frac{x^2}{2} + \frac{x^3}{3!} + \cdots + \frac{x^n}{n!} + \cdots$$

Lesson 4.06 Power Series

At the risk of driving you out of the lab, you are asked to look at this with a critical eye. To begin with, nobody has ever succeeded in giving a totally precise value for e itself. This means that if you are going to be totally precise, then you have to admit that you are not quite certain of what e^x actually means. This problem is compounded by the fact that even if you have a totally precise value for e, then computation of numbers such as e^π becomes a problem. Raising a number to the πth power does not have the same physical meaning that raising a number to the 3rd power or raising it to the $(1/4)$ power by taking its fourth root. In advanced mathematics, these problems are cleanly circumvented. In advanced mathematics, e^x is defined via the power series

$$1 + x + \frac{x^2}{2} + \frac{x^3}{3!} + \frac{x^4}{4!} + \cdots + \frac{x^n}{n!} + \cdots$$

You know this power series converges for $-\infty < x < \infty$; so from the advanced point of view, there is no lack of clarity or precision in the definition. Similarly in advanced mathematics, the trigonometric functions $\sin[x]$ and $\cos[x]$ are defined via the power series

$$\sin[x] = x - \frac{x^3}{3!} + \frac{x^5}{5!} - \frac{x^7}{7!} + \cdots + \frac{(-1)^n \, x^{2n+1}}{(2\,n+1)!} + \cdots$$

$$\cos[x] = 1 - \frac{x^2}{2!} + \frac{x^4}{4!} - \frac{x^6}{6!} + \cdots + \frac{(-1)^n \, x^{2n}}{(2\,n)!} + \cdots$$

These precise definitions allow you to throw away the compass, protractor, and ruler that are usually associated with the measurements associated with the trigonometric functions. These two definitions also make the mathematics easier too.

G.9.a) Give a new derivation of the formula $D[e^x, x] = e^x$ by differentiating its power series

$$1 + x + \frac{x^2}{2} + \frac{x^3}{3!} + \frac{x^4}{4!} + \cdots + \frac{x^k}{k!} + \cdots$$

and seeing what you get.

G.9.b) Give a new derivation of the formula $D[\sin[x], x] = \cos[x]$ by differentiating its power series

$$x - \frac{x^3}{3!} + \frac{x^5}{5!} - \frac{x^7}{7!} + \cdots + \frac{(-1)^n \, x^{2n+1}}{(2\,n+1)!} + \cdots$$

and seeing what you get.

■ G.10) Ratios and roots

This brain teaser deals with Euler's way of approximating roots of polynomials. In each case, get the expansion of the indicated function in powers of x through the x^{60} term. Put $r =$ (coefficient of x^{59})/(coefficient of x^{60}).

Then evaluate the numerical value in decimals of the denominator of the indicated function for $x = r$. Report anything of interest you observe.

G.10.a.i) $\quad \dfrac{1}{1 - x - x^2}$

G.10.a.ii) $\quad \dfrac{2x + 3}{1 - x - x^2}$

G.10.a.iii) $\quad \dfrac{1}{1 - x - x^2 - 2x^4}$

G.10.a.iv) $\quad \dfrac{x^2 + 3}{1 - x - x^2 - 2x^4}$

G.10.a.v) $\quad \dfrac{x^9 - 1}{1 - x - x^2 - 2x^5 + x^7}$

G.10.a.vi) $\quad \dfrac{1}{1 - 2x + 4x^2 - 3x^3}$

G.10.b.i) Assess your results.

G.10.b.ii) Try to formulate a theory that explains why this procedure will produce good approximations of roots of the denominator in some cases but is bound to fail in others.

LITERACY SHEETS

Approximations: Measuring Nearness

4.01 Splines Literacy Sheet

L.1) What is the order of contact between e^x and $1+x$ at $x=0$?

What is the order of contact between
$$e^x \text{ and } 1+x+\frac{x^2}{2} \quad \text{at } x=0?$$

What is the order of contact between
$$e^x \text{ and } 1+x+\frac{x^2}{2}+\frac{x^3}{3!} \quad \text{at } x=0?$$

L.2) What is the order of contact between
$$\sin[x] \text{ and } x-\frac{x^3}{3!} \quad \text{at } x=0?$$

L.3) What is the order of contact between
$$\cos[x] \text{ and } 1-\frac{x^2}{2} \quad \text{at } x=0?$$

What is the order of contact between
$$\cos[x] \text{ and } 1-\frac{x^2}{2}+\frac{x^4}{4!} \quad \text{at } x=0?$$

L.4) What is the order of contact between
$$\log[x] \text{ and } (x-1)-\frac{(x-1)^2}{2} \quad \text{at } x=1?$$

What is the order of contact between

$$\log[x] \text{ and } (x-1) - \frac{(x-1)^2}{2} + \frac{(x-1)^3}{3} \quad \text{at } x = 1?$$

L.5) You have two curves $y = f[x]$ and $y = g[x]$ with $f[a] = g[a]$ and you want to make a transition from the $f[x]$ curve to the $g[x]$ curve at $\{a, f[a]\} = \{a, g[a]\}$.

What kind of transition do you expect to get if

$$f'[a] \neq g'[a]?$$

L.6) You have two curves $y = f[x]$ and $y = g[x]$ with $f[a] = g[a]$ and you want to make a transition from the $f[x]$ curve to the $g[x]$ curve at

$$\{a, f[a]\} = \{a, g[a]\}.$$

What kind of transition do you expect to get if

$$f'[a] = g'[a], \quad \text{but} \quad f''[a] \neq g''[a]?$$

L.7) You have two curves $y = f[x]$ and $y = g[x]$ with $f[a] = g[a]$ and you want to make a transition from the $f[x]$ curve to the $g[x]$ curve at $\{a, f[a]\} = \{a, g[a]\}$. What kind of transition do you expect to get if

$$f'[a] = g'[a] \quad \text{and} \quad f''[a] = g''[a], \quad \text{but} \quad f'''[a] \neq g'''[a]?$$

L.8) You have two curves $y = f[x]$ and $y = g[x]$ with $f[a] = g[a]$ and you want to make a transition from the $f[x]$ curve to the $g[x]$ curve at

$$\{a, f[a]\} = \{a, g[a]\}.$$

What kind of transition do you expect to get if

$$f'[a] = g'[a],$$
$$f''[a] = g''[a],$$
$$f'''[a] = g'''[a]$$

and

$$f^{(4)}[a] = g^{(4)}[a],$$

but

$$f^{(5)}[a] \neq g^{(5)}[a]?$$

L.9) You have two curves $y = f[x]$ and $y = g[x]$ with $f[a] \neq g[a]$ and you want to make a transition from the $f[x]$ curve to the $g[x]$ curve at $\{a, f[a]\}$.

Can you make the transition without jumping?

L.10) If $f[x]$ and $g[x]$ have order of contact 2 at $x = a$, and you set

$$F[x] = \int_a^x f[t]\, dt \quad \text{and} \quad G[x] = \int_a^x g[t]\, dt,$$

then what order of contact do you expect $F[x]$ and $G[x]$ to have at $x = a$? What order of contact do you expect $f'[x]$ and $g'[x]$ to have at $x = a$?

L.11) Comment on the following statement:

"Order of contact and splines are important considerations if you want to design a railroad spur track, made so a train does not derail as it makes the transition from the main track to the spur."

L.12) Comment on the following statement, which is the key to many of the ideas in the upcoming lessons:

"When two functions $f[x]$ and $g[x]$ have a high order of contact at $x = 0$, then the transition from one curve to the other at $\{0, f[0]\} = \{0, g[0]\}$ is so smooth that the plots of $f[x]$ and $g[x]$ share a lot of ink as x advances from the left of 0 to the right of 0. In short, if you stay with x's close to 0, either curve could be used to get a reasonable approximation of the other."

4.02 Expansions in Powers of x Literacy Sheet

L.1) Write down the expansions of $1/(1-x)$, e^x, $\sin[x]$, and $\cos[x]$ in powers of x.

L.2) Write down the expansion of $1/(1-x)$ in powers of x and use it to write down the expansion of $1/(1+x^2)$ in powers of x.

L.3) Write down the expansion of e^x in powers of x and use it to write down the expansion of e^{-x^2} in powers of x.

L.4) Write down the expansion of $\sin[x]$ in powers of x and use it to write down the expansion of $\sin[2x]$ in powers of x.

L.5) Write down the expansion of $\cos[x]$ in powers of x and use it to write down the expansion of $\cos[x/2]$ in powers of x.

L.6) Write down the expansion of $1/(1-x)$ in powers of x and use it to come up with the expansion of $x^2/(1+x^5)$ in powers of x.

L.7) The derivative of $1/(1-x)$ is $1/(1-x)^2$. Write down the expansion of $1/(1-x)$ in powers of x and take its derivative to come up with the expansion of $1/(1-x)^2$ in powers of x.

L.8) Look at this:

```
In[1]:=
  Clear[f,x]; f[x_] = E^(-x) Cos[x];
  Normal[Series[f[x],{x,0,7}]]

Out[1]=
              3    4    5    7
             x    x    x    x
  1 - x +  ---  - --- + --- - ---
              3    6    30   630
```

How do you know that *Mathematica* is telling you that:

→ $e^{-x}\cos[x]$ and $1-x$ have order of contact 2 at $x=0$.

→ $e^{-x}\cos[x]$ and $1-x+\dfrac{x^3}{3}$ have order of contact 3 at $x=0$.

→ $e^{-x}\cos[x]$ and $1-x+\dfrac{x^3}{3}-\dfrac{x^4}{6}$ have order of contact 4 at $x=0$.

→ $e^{-x}\cos[x]$ and $1-x+\dfrac{x^3}{3}-\dfrac{x^4}{6}+\dfrac{x^5}{30}$ have order of contact 5 at $x=0$.

→ $e^{-x}\cos[x]$ and $1-x+\dfrac{x^3}{3}-\dfrac{x^4}{6}+\dfrac{x^5}{30}$ have order of contact 6 at $x=0$.

→ $e^{-x}\cos[x]$ and $1-x+\dfrac{x^3}{3}-\dfrac{x^4}{6}+\dfrac{x^5}{30}-\dfrac{x^7}{630}$ have order of contact 7 at $x=0$?

L.9) Quick! What is the expansion of $f[x]=1+x+x^2$ in powers of x?

L.10) What happens to the graphs of e^x and

$$1 + x + \frac{x^2}{2} + \frac{x^3}{3!} + \frac{x^4}{4!} + \cdots + \frac{x^n}{n!}$$

as you increase n?

What happens to the graphs of $\sin[x]$ and

$$x - \frac{x^3}{3!} + \frac{x^5}{5!} + \cdots + (-1)^n \frac{x^{2n+1}}{(2n+1)!}$$

as you increase n?

L.11) If you have a cheap calculator that only adds, subtracts, multiplies and divides, then what numbers would you enter to calculate a reasonably accurate numerical estimate of $e^{0.5}$?

What numbers would you enter to calculate a reasonably accurate numerical estimate for $\cos[0.5]$?

L.12) In what way is using more and more of the expansion of a given function $f[x]$ in powers of x like using more and more of the decimals of a given number?

L.13) How does the idea of order of contact begin to explain why you can think of the expansion of a function $f[x]$ in powers of x as a great storehouse of better and better approximations of $f[x]$?

L.14) Some folks like to say that if you take the expansion of a function $f[x]$ in powers of x and take the infinite sum of all of the terms in the expansion, then at least for x's near 0 you will hit $f[x]$ exactly. Where do you think that they got this notion?

4.03 Using Expansions Literacy Sheet

L.1) When you want to study the behavior of a function near a point $x = b$ other than $x = 0$, you can use expansions in powers of $(x - b)$.

What is the expansion of a function $f[x]$ in powers of $(x - b)$?

What does it have to do with order of contact?

L.2) Explain how to find the expansion of e^x in powers of $(x - 3)$.

L.3) Use expansions in powers of x to calculate the limits:

a. $\lim\limits_{x \to 0} \dfrac{\sin[x]}{x}$

b. $\lim\limits_{x \to 0} \dfrac{\sin[3\,x^2]}{x^2}$

c. $\lim\limits_{x \to 0} \dfrac{\sin[4\,x^4]}{\sin[2\,x^4]}$

d. $\lim\limits_{x \to 0} \dfrac{1 + x - e^x}{x^2}$

e. $\lim\limits_{x \to 0} \dfrac{1 - x - e^{-x}}{1 - \cos[x]}$

L.4) Here is the expansion of
$$f[x] = \sin[\tan[x]] - \tan[\sin[x]]$$
in powers of x through the x^9 term:

In[1]:=
```
Normal[Series[Sin[Tan[x]] - Tan[Sin[x]],{x,0,9}]]
```

Out[1]=
```
  7      9
-x     29 x
---  - -----
 30     756
```

And here is the expansion of
$$g[x] = \arcsin[\arctan[x]] - \arctan[\arcsin[x]]$$
in powers of x through the x^9 term:

In[2]:=
```
Normal[Series[ArcSin[ArcTan[x]] - ArcTan[ArcSin[x]], {x,0,9}]]
```

Out[2]=
```
  7      9
-x     13 x
---  + -----
 30     756
```

Use what you see to calculate
$$\lim_{x \to 0} \dfrac{f[x]}{g[x]}.$$

L.5) Write out the expansions in powers of x of e^{-x^2} and $\cos[3x]$ through the x^4 terms. Multiply the two partial expansions to obtain the expansion in powers of x of $e^{-x^2}\cos[3x]$ through the x^4 term. Throw away the higher order terms.

L.6) Going with $i = \sqrt{-1}$, what do you mean when you write e^{it}?

How do you get the formula
$$e^{-it} = \cos[t] - i\sin[t]$$
and where does the formula
$$\cos[t] = \frac{e^{it} + e^{-it}}{2}$$
come from?

Where does the formula
$$\cos[8t] = \frac{e^{i8t} + e^{-i8t}}{2}$$
come from?

L.7) If $f[x]$ approximates $g[x]$ within $\frac{1}{3000}$ for $10 \le x \le 12$, then $\int_{10}^{12} f[x]\,dx$ approximates $\int_{10}^{12} g[x]\,dx$ to how many accurate decimals?

L.8) Given a function $f[x]$ and a positive integer k, tell how to set a number c such that if $p[x]$ is another function with $|p[x] - f[x]| < c$ for $0 \le x \le 1/2$, then $\int_0^{1/2} f[x]\,dx = \int_0^{1/2} p[x]\,dx$ with an error of less than 10^{-k}.

Why is this of interest?

L.9) What part of the expansion of a function $f[x]$ in powers of x reflects the behavior of the function for x's close to 0?

L.10) Set a constant c so that the plots of $\sin[3x]$ and $1 - e^{cx}$ share lots of ink on small intervals centered at 0.

L.11) The only information you have on a certain function $f[x]$ is:

→ $f[x]$ changes when x changes and

→ $-1 \le f[x] \le 1$ for all x's between $-\infty$ and ∞.

Is it possible for a plot of a partial expansion of $f[x]$ to share ink with the plot of $f[x]$ all the way from $-\infty$ to $+\infty$? Why?

L.12) Here are plots of $f[x] = \sin[x]$ and $g[x] = x$ on an interval centered at $x = 0$:

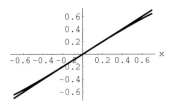

Use expansions to help explain why the plots turned out this way, thereby explaining why scientists often do not distinguish between $f[x] = \sin[x]$ and $g[x] = x$ when x is close to 0.

L.13) $\cos[x]$ is a transcendental function that is highly resistant to algebra. What can you do to coax a reasonable estimate of the number x^* with

$$\cos[x^*] - x^* = 0$$

from the machine?

L.14) What is a systematic way of doing Newton's Method for going after a good approximation to a true solution x^* of $f[x^*] = 0$ for any old function $f[x]$?

L.15) Newton's method gives a simple method for extracting square roots on a very cheap calculator. Here is the idea:

Given a number A, you try to calculate $\sqrt{A}$ by going with $f[x] = x^2 - A$ and using Newton's method to find an approximation of the number x^* that makes $f[x^*] = 0$.

Say what the update formula is, and then say how to punch the appropriate numbers into the cheap calculator.

L.16) If you've been around Calculus&Mathematica from the beginning, then you probably know that parabolic mirrors shaped like this concentrate vertical light rays at a single focal point:

```
In[3]:=
  Clear[x]
  Plot[-4 + (x^2)/3,{x,-2,2},
  PlotStyle->{{Blue,Thickness[0.01]}},
  AxesLabel->{"x",""},
  AspectRatio->Automatic];
```

Using $\sin[x]$ and x interchangeably for x's close to 0, the Sears, Zemansky, and Young freshman physics text, "University Physics" (Addison-Wesley, 1987, page 871), comes to the conclusion that some spherical mirrors also concentrate vertical light rays at a single focal point. They imply that a big spherical mirror like this will not concentrate vertical light rays at a single focal point:

240 Approximations: Measuring Nearness

In[4]:=
```
Plot[-Sqrt[4 - x^2],{x,-2,2},
PlotStyle->{{Blue,Thickness[0.01]}},
AxesLabel->{"x",""},
AspectRatio->Automatic];
```
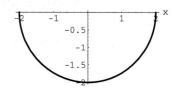

But they imply that this part of the spherical mirror will concentrate vertical light rays at a single focal point:

In[5]:=
```
Plot[-Sqrt[4 - x^2],{x,-0.4,0.4},
PlotStyle->{{Blue,Thickness[0.01]}},
PlotRange->{{-2,2},{-2,0}},
AxesLabel->{"x",""},
AspectRatio->Automatic];
```
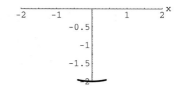

Here are the mathematical facts of the matter:

→ No part of a spherical mirror ever concentrates vertical light rays at a single focal point.

→ All parabolic mirrors like the one above do concentrate vertical light rays at a single focal point.

Look at the following:

In[6]:=
```
Clear[expan2,x]
expan2[x_] = Normal[Series[-Sqrt[4 - x^2],{x,0,2}]]
```
Out[6]=
$$-2 + \frac{x^2}{4}$$

This is the formula of a parabola. Now look at the following plots of the spherical mirror above and a closely related parabolic mirror:

In[7]:=
```
Plot[{-Sqrt[4 - x^2],expan2[x]},{x,-2,2},
PlotStyle->{{Thickness[0.02]},
{Thickness[0.01]}},PlotRange->{{-2,2},{-2,0}},
AxesLabel->{"x",""},
AspectRatio->Automatic];
```
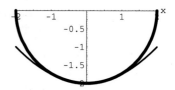

Eyeball this, and then identify the part of the spherical mirror that you believe will do a pretty good job of concentrating vertical light rays at a single point. Does this resolve the controversy between parabolic and spherical mirrors?

4.04 Taylor's Formula Literacy Sheet

L.1) All you know about a function $f[x]$ is:
$$f[0] = 2,$$
$$f'[0] = 6,$$
and
$$f''[0] = -8.$$
Write down the expansion of $f[x]$ in powers of x through the x^2 term.

L.2) All you know about a function $f[x]$ is:
$$f[2] = 1,$$
$$f'[2] = -3,$$
and
$$f''[2] = 1.$$
Write down the expansion of $f[x]$ in powers of $(x-2)$ through the $(x-2)^2$ term.

L.3) All you know about a pair of functions $f[x]$ and $g[x]$ is:
$$f[1] = 0,$$
$$f'[1] = 6,$$
$$g[1] = 0,$$
and
$$g'[1] = 2.$$
Calculate
$$\lim_{x \to 1} \frac{f[x]}{g[x]}.$$

L.4) All you know about a pair of functions $f[x]$ and $g[x]$ is:
$$f[1] = 0,$$
$$f'[1] = 0,$$
$$f''[1] = 8,$$
$$g[1] = 0,$$
$$g'[1] = 0,$$
and
$$g''[1] = 2.$$
Calculate
$$\lim_{x \to 1} \frac{f[x]}{g[x]}.$$

L.5) Here is the expansion of $f[x] = e^{\sin[\pi x]}$ in powers of $x - 1$ through the $(x-1)^4$ term:

In[1]:=
```
Normal[Series[E^(Sin[Pi x]),{x,1,4}]]
```

Out[1]=
$$1 - \text{Pi}\,(-1 + x) + \frac{\text{Pi}^2\,(-1+x)^2}{2} - \frac{\text{Pi}^4\,(-1+x)^4}{8}$$

Here is Taylor's formula for the expansion of a cleared function $f[x]$ in powers of $(x - 1)$ through the $(x - 1)^4$ term:

In[2]:=
```
Clear[f]; Normal[Series[f[x],{x,1,4}]]
```

Out[2]=
$$f[1] + (-1+x)\,f'[1] + \frac{(-1+x)^2\,f''[1]}{2} + \frac{(-1+x)^3\,f^{(3)}[1]}{6} + \frac{(-1+x)^4\,f^{(4)}[1]}{24}$$

Use what you see above to write down the values of

$$f'[1],\ f''[1],\ f^{(3)}[1] = f'''[1],\text{ and } f^{(4)}[1] = f''''[1]$$

in the case that $f[x] = e^{\sin[\pi x]}$.

L.6) Here is the expansion of

$$f[x] = \sin[\tan[x]] - \tan[\sin[x]]$$

in powers of x through the x^9 term:

In[3]:=
```
Normal[Series[Sin[Tan[x]] - Tan[Sin[x]],{x,0,9}]]
```

Out[3]=
$$-\frac{x^7}{30} - \frac{29\,x^9}{756}$$

No, that's not a misprint. How does this reveal that

$$f[0] = f'[0] = f''[0] = f'''[0] = f^{(4)}[0] = f^{(5)}[0] = f^{(6)}[0] = 0$$

but $f^{(7)}[0] \neq 0$?

L.7) Take two functions $f[x]$ and $g[x]$ such that

→ $f[b] = 0$

→ $g[b] = 0$ and

→ $g'[b] \neq 0$

and look at:

In[4]:=
```
Clear[f,g,x,b]; Normal[Series[f[x],{x,b,2}]]/.f[b]->0
```
Out[4]=
$$(-b + x)\, f'[b] + \frac{(-b + x)^2\, f''[b]}{2}$$

In[5]:=
```
Normal[Series[g[x],{x,b,2}]]/.g[b]->0
```
Out[5]=
$$(-b + x)\, g'[b] + \frac{(-b + x)^2\, g''[b]}{2}$$

How does this tell you that
$$\lim_{x \to b} \frac{f[x]}{g[x]} = \frac{f'[b]}{g'[b]}?$$

Some folks call this formula L'Hospital's Rule.

L.8) Take two functions $f[x]$ and $g[x]$ such that

→ $f[b] = 0$ and $g[b] = 0$
→ $f'[b] = 0$ and $g'[b] = 0$ and
→ $g''[b] \neq 0$

and look at:

In[6]:=
```
Clear[f,g,x,b]
Normal[Series[f[x],{x,b,3}]]/.{f[b]->0,f'[b]->0}
```
Out[6]=
$$\frac{(-b + x)^2\, f''[b]}{2} + \frac{(-b + x)^3\, f^{(3)}[b]}{6}$$

In[7]:=
```
Normal[Series[g[x],{x,b,3}]]/.{g[b]->0,g'[b]->0}
```
Out[7]=
$$\frac{(-b + x)^2\, g''[b]}{2} + \frac{(-b + x)^3\, g^{(3)}[b]}{6}$$

How does this tell you that
$$\lim_{x \to b} \frac{f[x]}{g[x]} = \frac{f''[b]}{g''[b]}?$$

Some folks call this formula L'Hospital's Rule.

L.9) Take two functions $f[x]$ and $g[x]$ such that

- $f[b] = 0$ and $g[b] = 0$
- $f'[b] = 0$ and $g'[b] = 0$
- $f''[b] = 0$ and $g''[b] = 0$ and
- $g'''[b] \neq 0$

Give a formula for

$$\lim_{x \to b} \frac{f[x]}{g[x]}$$

and explain where you got it. Some folks call this formula L'Hospital's Rule.

L.10) Professor Lipman Bers of Columbia University, and past president of the American Mathematical Society, once said: "L'Hospital's rule impressed the early practitioners of calculus. The importance of this method in contemporary mathematics is not overwhelming." Why is L'Hospital's rule no big deal if you know Taylor's formula?

L.11) Which of the following do you regard as the best overall approximation on $[-1, 1]$ of $f[x] = e^x$:

```
approx1[x] = Normal[Series[E^x,{x,-1,8}]]
approx2[x] = Normal[Series[E^x,{x,0,8}]]
approx3[x] = Normal[Series[E^x,{x,1,8}]]?
```

How would you use your selection to estimate the value of $\int_0^1 e^{-x^4}\, dx$?

L.12) Write down the expansions of $1/(1-x)$ and e^x in powers of x.

L.13) Here is an expansion in powers of x of $f[x] = (x+1)\sin[x^2]$ through the x^6 term:

In[8]:=
```
Normal[Series[(x+1)Sin[x^2],{x,0,6}]]
```

Out[8]=

$$x^2 + x^3 - \frac{x^6}{6}$$

Use what you see to produce the table of values:

$$\{f[0],\ f'[0],\ f''[0],\ f'''[0],\ f^{(4)}[0],\ f^{(5)}[0],\ f^{(6)}[0]\}.$$

L.14) Here is a table of $f[1]$ together with the first four derivatives: $f'[1], f''[1], f'''[1]$, and $f^{(4)}[1]$, for $f[x] = e^{\cos[x]}$:

In[9]:=
```
Clear[f,x]; f[x_]=Exp[Cos[x]]; Table[D[f[x],{x,k}]/.x->1.,{k,0,4}]
```

Out[9]=
```
{1.71653, -1.44441, 0.287983, 2.76291, -5.51054}
```

Use what you see to give an expansion of $f[x]$ in powers of $(x-1)$ through the $(x-1)^4$ term.

L.15) Sometimes you have a formula for $f'[x]$ and the value of $f[b]$. But in cases like $f'[x] = e^{\cos[x]}$, the formula for $f[x]$ is not available. In order to fake the plot of the $f[x]$ for x's near b you have several choices:

Euler approximation:

In[10]:=
```
Clear[feuler,f,x,b]; feuler[x_,b_] = f[b] + (x - b) f'[b]
```

Out[10]=
```
f[b] + (-b + x) f'[b]
```

Check the order of contact at $x = b$:

In[11]:=
```
{Series[feuler[x,b],{x,b,2}],Series[f[x],{x,b,2}]}
```

Out[11]=
$$\{f[b] + f'[b] (-b + x) + O[-b + x]^3,$$
$$f[b] + f'[b] (-b + x) + \frac{f''[b] (-b + x)^2}{2} + O[-b + x]^3\}$$

a) What is the order of contact between $f[x]$ and feuler$[x, b]$ at $x = b$?

Trapezoidal approximation:

In[12]:=
```
Clear[ftrap,f,x,b]
ftrap[x_,b_] = f[b] + (x - b) (f'[b] + f'[x])/2
```

Out[12]=
$$f[b] + \frac{(-b + x) (f'[b] + f'[x])}{2}$$

Check the order of contact at $x = b$:

In[13]:=
```
{Series[ftrap[x,b],{x,b,3}],Series[f[x],{x,b,3}]}
```

Out[13]=
$$\{f[b] + f'[b] (-b + x) + \frac{f''[b] (-b + x)^2}{2} + \frac{f^{(3)}[b] (-b + x)^3}{4} + O[-b + x]^4,$$
$$f[b] + f'[b] (-b + x) + \frac{f''[b] (-b + x)^2}{2} + \frac{f^{(3)}[b] (-b + x)^3}{6} + O[-b + x]^4\}$$

b) What is the order of contact between $f[x]$ and ftrap$[x,b]$ at $x = b$?

Midpoint approximation:

In[14]:=
```
Clear[fmidpt,f,x,b]
fmidpt[x_,b_] = f[b] + (x - b) f'[(b + x)/2]
```

Out[14]=
$$f[b] + (-b + x)\, f'\!\left[\frac{b + x}{2}\right]$$

Check the order of contact at $x = b$:

In[15]:=
```
{Series[fmidpt[x,b],{x,b,3}],Series[f[x],{x,b,3}]}
```

Out[15]=
$$\left\{f[b] + f'[b](-b+x) + \frac{f''[b](-b+x)^2}{2} + \frac{f^{(3)}[b](-b+x)^3}{8} + O[-b+x]^4,\right.$$

$$\left. f[b] + f'[b](-b+x) + \frac{f''[b](-b+x)^2}{2} + \frac{f^{(3)}[b](-b+x)^3}{6} + O[-b+x]^4\right\}$$

c) What is the order of contact between $f[x]$ and ftmidpt$[x,b]$ at $x = b$?

Which comes closer to having order of contact 3 with $f[x]$ at $x = b$: ftrap$[x,b]$ or fmidpt$[x,b]$?

Runge-Kutta approximation:

In[16]:=
```
Clear[frunge,f,x,b]
frunge[x_,b_] = (2 fmidpt[x,b] + ftrap[x,b])/3
```

Out[16]=
$$f[b] + \frac{\dfrac{(-b+x)(f'[b] + f'[x])}{2} + 2\left(f[b] + (-b+x)\,f'\!\left[\dfrac{b+x}{2}\right]\right)}{3}$$

Check the order of contact at $x = b$:

In[17]:=
```
{Series[frunge[x,b],{x,b,5}],Series[f[x],{x,b,5}]}
```

Out[17]=
$$\left\{f[b] + f'[b](-b+x) + \frac{f''[b](-b+x)^2}{2} + \frac{f^{(3)}[b](-b+x)^3}{6} + \right.$$

$$\left. \frac{f^{(4)}[b](-b+x)^4}{24} + \frac{5\,f^{(5)}[b](-b+x)^5}{576} + O[-b+x]^6,\right.$$

$$f[b] + f'[b](-b+x) + \frac{f''[b](-b+x)^2}{2} + \frac{f^{(3)}[b](-b+x)^3}{6} +$$

$$\frac{f^{(4)}[b](-b+x)^4}{24} + \frac{f^{(5)}[b](-b+x)^5}{120} + O[-b+x]^6\}$$

d) What is the order of contact between $f[x]$ and frunge$[x, b]$ at $x = b$?

e) Why do some folks say that even though the order of contact between $f[x]$ and frunge$[x, b]$ at $x = b$ is not 5, for practical purposes, the order of contact at $x = b$ is almost 5?

f) In serious heavy calculations, why would most professional number crunchers prefer to approximate $f[x]$ near $x = b$ by frunge$[x, b]$ instead of:

In[18]:=
  Normal[Series[f[x],{x,b,5}]]

Out[18]=

$$f[b] + (-b+x)f'[b] + \frac{(-b+x)^2 f''[b]}{2} + \frac{(-b+x)^3 f^{(3)}[b]}{6} +$$

$$\frac{(-b+x)^4 f^{(4)}[b]}{24} + \frac{(-b+x)^5 f^{(5)}[b]}{120}$$

4.05 Barriers to Convergence Literacy Sheet

L.1) At what point on the x-axis are the convergence intervals of expansions in powers of $(x-1)$ centered?

At what point on the x-axis are the convergence intervals of expansions in powers of $(x+5)$ centered?

L.2) Use complex numbers and singularities to explain why, when you expand
$$f[x] = \frac{1}{1+x^2}$$
in powers of x, you run into barriers at $x=-1$ and $x=1$.

L.3) Use complex numbers and singularities to explain why, when you expand
$$f[x] = \arctan[x]$$
in powers of x, you run into barriers at $x=-1$ and $x=1$.

L.4) Use complex numbers and singularities to explain why when you try to expand
$$f[x] = x^{5/2}$$
in powers of x, you must fail.

L.5) Give the convergence interval for the expansions of the following functions $f[x]$ in powers of $(x-a)$ for the given choices of a:

- $f[x] = \dfrac{1}{1-x}$ with $a = 0$
- $f[x] = \dfrac{1}{1-x}$ with $a = 4$
- $f[x] = e^x$ with $a = 0$
- $f[x] = \sin[x]$ with $a = 0$
- $f[x] = \sin[x]$ with $a = \pi$
- $f[x] = \cos[x]$ with $a = 0$
- $f[x] = \sqrt{1+x}$ with $a = 0$
- $f[x] = x^{5/3}$ with $a = 1$
- $f[x] = \arctan[x]$ with $a = 0$ and
- $f[x] = \dfrac{1}{1+x^2}$ with $a = 1$.

L.6) The expansion of $1/(1+x^2)$ in powers if x is
$$1 - x^2 + x^4 - x^6 + x^8 + \cdots + (-1)^k x^{2k} + \cdots.$$

Explain the following statement:

Although there is nothing wrong with saying

$$\frac{4}{5} = \frac{1}{1+1/2^2} = 1 - \left(\frac{1}{2}\right)^2 + \left(\frac{1}{2}\right)^4 - \left(\frac{1}{2}\right)^6 + \left(\frac{1}{2}\right)^8 + \cdots + (-1)^k \left(\frac{1}{2}\right)^{2k} + \cdots,$$

it would be a sure sign of calculus illiteracy to say

$$\frac{1}{5} = \frac{1}{1+2^2} = 1 - 2^2 + 2^4 - 2^6 + 2^8 + \cdots + (-1)^k 2^{2k} + \cdots.$$

L.7) Write down the number that is

$$1 + \frac{1}{2} + \left(\frac{1}{2}\right)^2 + \left(\frac{1}{2}\right)^3 + \cdots + \left(\frac{1}{2}\right)^k + \cdots.$$

Write down the number that is

$$\frac{1}{2} + \left(\frac{1}{2}\right)^2 + \left(\frac{1}{2}\right)^3 + \cdots + \left(\frac{1}{2}\right)^k + \cdots.$$

Write down the number that is

$$1 + \frac{1}{2} + \frac{1}{2!\, 3^2} + \frac{1}{3!\, 3^3} + \cdots + \frac{1}{k!\, 3^k} + \cdots.$$

Write down the number that is

$$1 - \frac{\pi^2}{2!} + \frac{\pi^4}{4!} + \cdots + \frac{(-1)^n \pi^{2n}}{(2n)!} + \cdots.$$

Write down the number that is

$$\frac{1}{2} - \left(\frac{1}{2}\right)^2 + \left(\frac{1}{2}\right)^3 - \left(\frac{1}{2}\right)^4 + \cdots + (-1)^{n+1} \left(\frac{1}{2}\right)^n + \cdots.$$

Write down the number that is

$$\frac{\pi}{2} - \frac{\pi^3}{2^3\, 3!} + \frac{\pi^5}{2^5\, 5!} + \cdots + \frac{(-1)^k \pi^{2k+1}}{2^{2k+1}\, (2k+1)!} + \cdots.$$

L.8) The government injects $90 billion into the economy. Each recipient spends 90% of the dollars received. In turn, the secondary recipients spend 90% of the dollars they receive, etc.

Use the expansion of $1/(1-x)$ in powers of x to come up with a clean measurement of the total spending that results from the $90 billion injection.

L.9) The government injects $100 billion into the economy. But this time each recipient spends only 50% of the dollars received. In turn, the secondary recipients spend 50% of the dollars they receive, etc.

Use the expansion of $1/(1-x)$ in powers of x to come up with a clean measurement of the total spending that results from the $100 billion injection.

L.10) Neither $\sqrt{x}$ nor $\log[x]$ has an expansion in powers of x. But if $a > 0$, then both functions have expansions in powers of $(x - a)$. How do you account for this?

L.11) Why is it probably a good idea to approximate $e^{0.5}$ by
$$1 + 0.5 + \frac{(0.5)^2}{2!} + \frac{(0.5)^3}{3!} + \frac{(0.5)^4}{4!} + \frac{(0.5)^5}{5!}?$$
And why is it probably a bad idea to approximate e^{20} by
$$1 + 20 + \frac{(20)^2}{2!} + \frac{(20)^3}{3!} + \frac{(20)^4}{4!} + \frac{(20)^5}{5!}?$$

L.12) Agree or disagree, and explain yourself:

Because e^x has no complex singularities, the expansion of e^x in powers of x converges to e^x no matter what x is.

L.13) Agree or disagree, and explain yourself:

The fact that the expansion of e^x in powers of x converges to e^x no matter what x is of limited practical value because, when x is very large, the expansion converges too slowly to be of any use in getting a precise calculation of e^x.

L.14) The decimal form of a number is useful because it gives you an easy way of understanding the location and size of numbers. Some numbers have a decimal form that stops after a few digits:

In[1]:=
N[1453/8,20]

Out[1]=
181.625

Others are not so obliging:

In[2]:=
N[20/6]

Out[2]=
3.33333

In[3]:=
N[20/6,20]

Out[3]=
3.3333333333333333333

Sometimes a pattern is discernible:

In[4]:=
N[22/7,7]

Out[4]=
3.142857

In[5]:=
N[22/7,13]

Out[5]=
   3.142857142857

In[6]:=
   N[22/7,19]

Out[6]=
   3.142857142857142857

Sometimes no pattern is discernible:

In[7]:=
   N[Pi,18]

Out[7]=
   3.14159265358979324

In[8]:=
   N[Pi,60]

Out[8]=
   3.14159265358979323846264338327950288419716939937510582097494

An analogous situation occurs in the realm of functions. Some functions have an expansion in powers of x that stops after a few terms:

In[9]:=
   Clear[x,f]
   f[x_] = 1 - 2 x + 3 x^2 + 4 x^4 - 23 x^10;
   Normal[Series[f[x],{x,0,4}]]

Out[9]=
$$1 - 2x + 3x^2 + 4x^4$$

In[10]:=
   Normal[Series[f[x],{x,0,9}]]

Out[10]=
$$1 - 2x + 3x^2 + 4x^4$$

In[11]:=
   Normal[Series[f[x],{x,0,12}]]

Out[11]=
$$1 - 2x + 3x^2 + 4x^4 - 23x^{10}$$

In[12]:=
   Normal[Series[f[x],{x,0,100}]]

Out[12]=
$$1 - 2x + 3x^2 + 4x^4 - 23x^{10}$$

Others are not so obliging:

In[13]:=
   Clear[x,f]; f[x_] = 1/(1 + x^2);
   Normal[Series[f[x],{x,0,4}]]

Out[13]=
$$1 - x^2 + x^4$$

In[14]:=
```
Normal[Series[f[x],{x,0,14}]]
```

Out[14]=
$$1 - x^2 + x^4 - x^6 + x^8 - x^{10} + x^{12} - x^{14}$$

Sometimes a pattern is discernible:

In[15]:=
```
Clear[x,f]; f[x_] = 1/(1 - x)^2;
Normal[Series[f[x],{x,0,4}]]
```

Out[15]=
$$1 + 2x + 3x^2 + 4x^3 + 5x^4$$

In[16]:=
```
Normal[Series[f[x],{x,0,12}]]
```

Out[16]=
$$1 + 2x + 3x^2 + 4x^3 + 5x^4 + 6x^5 + 7x^6 + 8x^7 + 9x^8 + 10x^9 + 11x^{10} + 12x^{11} + 13x^{12}$$

Sometimes no pattern is discernible:

In[17]:=
```
Clear[x,f]; f[x_] = Tan[x];
Normal[Series[f[x],{x,0,3}]]
```

Out[17]=
$$x + \frac{x^3}{3}$$

In[18]:=
```
Normal[Series[f[x],{x,0,7}]]
```

Out[18]=
$$x + \frac{x^3}{3} + \frac{2x^5}{15} + \frac{17x^7}{315}$$

In[19]:=
```
Normal[Series[f[x],{x,0,13}]]
```

Out[19]=
$$x + \frac{x^3}{3} + \frac{2x^5}{15} + \frac{17x^7}{315} + \frac{62x^9}{2835} + \frac{1382x^{11}}{155925} + \frac{21844x^{13}}{6081075}$$

Just as decimals are useful in allowing you to understand the size and exact location of various numbers, this form has proven to be a good form to allow the human mind to grasp the true nature of functions. Use the theme of approximation to react in words to the sentiments and calculations above.

4.06 Power Series Literacy Sheet

L.1) Agree or disagree:

Every expansion of a function in powers of x is a power series.

Agree or disagree:

Every expansion of a function in powers of $(x-1)$ is a power series.

L.2) Agree or disagree:

Some functions don't have clean, simple formulas, so instead of defining them by formulas, these weirdo functions are sometimes defined by power series.

L.3) What does it mean to say that a function is defined by a power series?

L.4) When you are given a function that is defined by the power series that is the function's expansion in powers of x, what can you learn about the function from the power series?

How much of the plot can you expect to get?

L.5) Draw hand sketches of

$$y = x \quad \text{and} \quad y = x + x^{10} + x^{12} + x^{100} + x^{102} \quad \text{for } -1.2 \le x \le 1.2$$

on these axes.

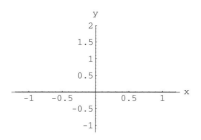

What relevance does your plot have when you are talking about the general topic of plots of functions that are defined by power series?

L.6) Sometimes, a function is defined as the solution of a differential equation. For instance, you can define $\text{lit}[x]$ to be the solution of

$$\text{lit}'[x] = x^2 \, \text{lit}[x] \left(1 - \frac{\text{lit}[x]}{8}\right) - 3 \quad \text{with } \text{lit}[0] = 1.$$

What rote procedure can you invoke to determine the power series that is the expansion of $\text{lit}[x]$ in powers of x?

L.7) For the following differential equations, give guaranteed convergence intervals for the expansion of the solutions in powers of x.

→ $y''[x] - y[x] = \sin[x]$ with $y[0] = 3$ and $y'[0] = -1$.

→ $(1-x)\,y''[x] - y[x] = \sin[x]$ with $y[0] = 3$ and $y'[0] = -1$.

→ $y''[x] + y'[x] - 4\,y[x] = \dfrac{1}{1+4\,x^2}$ with $y[0] = 0$ and $y'[0] = 1$.

L.8) Find convergence intervals of the following power series. Use the Power Series Convergence Principle directly, use the Ratio Test, or use any other method you like.

→ $1 + x + x^2 + x^3 + x^4 + x^5 + \cdots + x^n + \cdots$

→ $1 + 2x + 4x^2 + 8x^3 + 16x^4 + 32x^5 + \cdots + 2^n x^n + \cdots$

→ $1 + x + \dfrac{x^2}{5^2} + \dfrac{x^3}{5^3} + \dfrac{x^4}{5^4} + \cdots + \dfrac{x^n}{5^n} + \cdots$

→ $1 + x + \dfrac{x^2}{2} + \dfrac{x^3}{3} + \dfrac{x^4}{4} + \cdots + \dfrac{x^n}{n} + \cdots$

→ $1 + x + \dfrac{x^2}{2^{1/2}} + \dfrac{x^3}{3^{1/3}} + \dfrac{x^5}{5^{1/5}} + \cdots + \dfrac{x^k}{k^{1/k}} + \cdots$

Heavy Tip: $k^{1/k} > 1$ for $k > 1$

L.9) When you realize that $y[x] = e^x$ is the solution of the differential equation

$y''[x] - y[x] = 0$ with $y[0] = 1$ and $y'[0] = 1$,

then you know automatically that the expansion

$1 + x + \dfrac{x^2}{2} + \dfrac{x^3}{3!} + \cdots + \dfrac{x^k}{k!} + \cdots$

converges to e^x no matter what x is. Why is this automatic?

L.10) Suppose

$$f[x] = 1 + x + \dfrac{x^2}{2^2} + \dfrac{x^3}{3^2} + \cdots + \dfrac{x^k}{k^2} + \cdots$$

→ What is the value of $f[0]$?

→ What is the value of $f'[0]$?

→ What is the value of $f''[0]$?

→ What is the value of $f^{(5)}[0]$?

L.11) Is the infinite sum
$$1 + \frac{1}{2} + \frac{1}{3} + \frac{1}{4} + \frac{1}{5} + \frac{1}{6} + \cdots + \frac{1}{n} + \cdots$$
convergent?

L.12) Discuss:

  a. The role of the expansion of $1/(1-t)$ in powers of t in the explanation of the Basic Convergence Principle.

  b. The role of the Basic Convergence Principle in the explanation of the Ratio Test.

Index

Entries are listed by lesson number and problem number.

Airplanes, landing, splines and, 4.01: T.2
Airy function, power series and, 4.06: G.6
Approximations
  based on Taylor's formula, 4.04: B.2
  of e^x, sin[x] and cos[x], barriers to convergence and, 4.05: B.2
  by expansions in powers of $(x - b)$, 4.03: G.1
  expansions in powers of x for, 4.02: B.3, G.1, G.4
  midpoint and Runge-Kutta, expansions versus, Taylor's formula and, 4.04: G.6
  midpoint versus trapezoidal, Taylor's formula and, 4.04: G.7
  of solutions of differential equations, Taylor's formula and, 4.04: T.3, G.8
  Taylor's formula and. *See* Taylor's formula

Barriers to convergence
  $\int_t^a$ and log[x], 4.05: G.9
  Calculus Cal and, 4.05: G.8
  complex numbers and, 4.05: B.1
  convergence intervals and, 4.05: T.1, G.1, G.2
  for $f[x]$, $f'[x]$ and $\int_a^t f[t]dt$, 4.05: T.2
  interpolating, 4.05: G.12
  plots and, 4.05: G.2
  convergence levels and, 4.05: B.4
  drug dosing and, 4.05: G.6
  expansion of $1/(1-x)$ for, 4.05: T.5

  expansions and
    infinite sums and decimals and, 4.05: G.11
    interpolating, 4.05: G.12
    of $\int_a^\infty = 1 + x + x^2 + x^3 + x^4 + \cdots$ and, 4.05: G.7
  functions without expansions in powers of x and, 4.05: B.3
  impossible and outrageous problems and, 4.05: G.10
  infinite sums and, 4.05: T.4, G.4
    expansions and, 4.05: G.11
  lack of, with approximations of e^x, sin[x] and cos[x], 4.05: B.2
  polynomials and, interpolating, 4.05: G.12
  resulting from splines, 4.05: G.5
  sharing ink and, 4.05: G.3
  $1/(1-x) = 1 + x + x^2 + \cdots + x^k + \cdots$ for $-1 < x < 1$, 4.05: T.3

Calculus Cal, barriers to convergence and, 4.05: G.8
Centering expansions, 4.03: G.3
Circles, expansions in powers of x and, 4.02: G.5
Complex exponential function
  expansions and, 4.03: B.5
  generating trigonometric identities using, 4.03: T.3
Complex numbers, barriers to convergence and, 4.05: B.1

259

Index

Convergence. *See* Barriers to convergence; Convergence intervals; Convergence levels
Convergence intervals
   barriers to convergence and, 4.05: T.1, G.1
      for $f[x]$, $f'[x]$ and $\int_a^t f[t]\,dt$, 4.05: T.2
   interpolating intervals and, 4.05: G.12
   plots and, 4.05: G.2
   for functions defined by power series, via differential equations, 4.06: B.3, T.5
   interpolating, barriers to convergence and, 4.05: G.12
   for power series
      general, 4.06: B.4
      guaranteed, 4.06: G.5
Convergence levels, barriers to convergence and, 4.05: B.4
Cubic splines, natural, 4.01: G.6

Decimals, barriers to convergence and, 4.05: G.11
Differential equations
   approximations of solutions of, Taylor's formula and, 4.04: G.8
   functions defined by power series and, 4.06: B.2, T.2, G.3
      convergence intervals for, 4.06: B.3, T.5
Differentiation, expansions in powers of x by, 4.02: T.2
Division, expansions of $\tan[x]$ by, 4.03: G.7
Drug dosing, barriers to convergence and, 4.05: G.6
expansion of $1/(1-x)$ for, 4.05: T.5

Expansions
   barriers to convergence and, 4.05: G.11
      expansions of $\int_a^\infty = 1 + x + x^2 + x^3 + x^4 + \cdots$ and, 4.05: G.7
   interpolating, 4.05: G.12
   behavior away from 0, 4.03: G.10
   behavior close to 0, 4.03: G.9
   centering, 4.03: G.3
   complex exponential function and, 4.03: B.5
      generating trigonometric identities using, 4.03: T.3
   of derivatives, Taylor's formula and, 4.04: G.3
   error at endpoints and, 4.03: G.11
   integral estimation using, 4.03: T.4, G.5
   interpolating, barriers to convergence and, 4.05: G.12
   limit calculation using, 4.03: B.4, T.1, G.2
   midpoint and Runge-Kutta approximations versus, Taylor's formula and, 4.04: G.6
   Newton's method and, tangent lines and, 4.03: B.2

parabolic mirrors versus spherical mirrors and, 4.03: G.6
in powers of $(x-b)$, 4.03: B.1
   approximation of functions by, 4.03: G.1
in powers of x, 4.02: B.1, B.2, G.6
   for approximation, 4.02: B.3, G.1, G.4
   circles and, 4.02: G.5
   by differentiation, 4.02: T.2
   by integration, 4.02: T.3, G.7
   by substitution, 4.02: T.1
   trick questions and, 4.02: G.3
   writing, 4.02: G.2
Series instruction and, 4.03: B.3
$\sin[mx]$ and, generating identities for, 4.03: G.8
spherical mirrors versus parabolic mirrors and, 4.03: G.6
square roots by Newton's method and, 4.03: T.2, G.4
tangent lines and Newton's method and, 4.03: B.2
of $\tan[x]$, by division, 4.03: G.7
Taylor's formula and
   of $1/(1-x)$, $\sin[x]$, $\cos[x]$ and e^x in powers of $(x-b)$, 4.04: G.4
   expansions of derivatives and, 4.04: G.3
trigonometric identity generation using, complex exponential function and, 4.03: T.3
of $\int_a^\infty = 1 + x + x^2 + x^3 + x^4 + \cdots$ and, barriers to convergence and, 4.05: G.7

Fake plots
   of solutions of differential equations, Taylor's formula and, 4.04: G.8
   of solutions of differential equations and, Taylor's formula and, 4.04: T.3

Infinite sums
   barriers to convergence and, 4.05: T.4, G.4, G.11
   power series and, 4.06: T.4, G.7
Integrals, estimating using expansions, 4.03: T.4, G.5
Integration, expansions in powers of x by, 4.02: T.3, G.7
Interpolation function and, smooth splines and, 4.01: T.3

Kissing parabolas, Taylor's formula and, 4.04: G.9

Limits
   calculating using expansions, 4.03: B.4, T.1, G.2

Limits (*continued*)
    Taylor's formula and, 4.04: T.2
    L'Hospital's rule and, 4.04: G.2

Midpoints, Taylor's formula and, 4.04: G.5
    approximations and, 4.04: G.7
    midpoint and Runge-Kutta approximations versus expansions and, 4.04: G.6
    midpoint versus trapezoidal approximations and, 4.04: G.7

Natural cubic splines, 4.01: G.6
Newton's method
    square roots by, 4.03: T.2, G.4
    tangent lines and, 4.03: B.2

Parabolas, Taylor's formula and, 4.04: G.5
    kissing parabolas and, 4.04: G.9
Parabolic mirrors, controversy between spherical mirrors and, expansions and, 4.03: G.6
Polynomials
    interpolating, barriers to convergence and, 4.05: G.12
    splines and, 4.01: T.1, G.2
Power series
    advanced viewpoint of e, $\sin[x]$, and $\cos[x]$ and, 4.06: G.9
    convergence intervals and. *See* Convergence intervals
    functions defined by, 4.06: B.1
        plotting, 4.06: T.1, G.1, G.8
    via differential equations, 4.06: B.2, B.3, T.2, T.5, G.3
    infinite sums of numbers and, 4.06: T.4, G.7
    ratios and roots and, 4.06: G.10
    ratio test and, 4.06: T.3
    simple Airy function and, 4.06: G.6
    Taylor's formula and, 4.06: G.2

Ratios, power series and, 4.06: T.3, G.10
Rectangles, Taylor's formula and, 4.04: G.5
Road design, splines and, 4.01: G.3
Roots, power series and, 4.06: G.10
Runge-Kutta approximation, expansions and, Taylor's formula and, 4.04: G.6

Series instruction, expansions and, 4.03: B.3
Sharing ink, barriers to convergence and, 4.05: G.3
$\sin[mx]$ and, generating identities for, expansions and, 4.03: G.8
Spherical mirrors, controversy between parabolic mirrors and, expansions and, 4.03: G.6
Splines
    barriers to convergence resulting from, 4.05: G.5
    landing airplanes and, 4.01: T.2
    natural cubic, 4.01: G.6
    order of contact and
        changing variables to improve order of contact at 0, 4.01: G.5
        for derivatives and integrals, 4.01: G.4
        plots and, 4.01: B.1
    plots and, 4.01: G.1
        order of contact and, 4.01: B.1
    polynomials and, 4.01: T.1, G.2
    road design and, 4.01: G.3
    smooth, 4.01: B.2
        Interpolation function and, 4.01: T.3
Square roots, by Newton's method, 4.03: T.2, G.4
Substitution, expansions in powers of x by, 4.02: T.1
Sums, infinite
    barriers to convergence and, 4.05: T.4, G.4, G.11
    power series and, 4.06: T.4, G.7

Tangent lines, Newton's method and, 4.03: B.2
$\tan[x]$, expansions by division, 4.03: G.7
Taylor's formula, 4.04: B.1
    approximations and, 4.04: B.2
    midpoint and Runge-Kutta, expansions versus, 4.04: G.6
    midpoint versus trapezoidal, 4.04: G.7
    of solutions of differential equations, 4.04: G.8
    solutions of differential equations and, 4.04: T.3
    expansions and
        of $1/(1-x)$, $\sin[x]$, $\cos[x]$ and e^x in powers of $(x-b)$, 4.04: G.4
        of derivatives, 4.04: G.3
    fake plots and, of solutions of differential equations, 4.04: T.3, G.8
    kissing parabola and, 4.04: G.9
    limits and, 4.04: T.2
        L'Hospital's rule and, 4.04: G.2
    power series and, 4.06: G.2
    rectangles, trapezoids, midpoints, and parabolas and, 4.04: G.5
    in reverse, 4.04: T.1, G.1
Trapezoids, Taylor's formula and, 4.04: G.5
    approximations and, 4.04: G.7